The Physical Geography, Geomorphology and Late Quaternary History of the Mahidasht Project Area, Qara Su Basin Central West Iran

ROM Mahidasht Project
Volume 1

Louis D. Levine, General Editor

The Physical Geography, Geomorphology and Late Quaternary History of the Mahidasht Project Area, Qara Su Basin Central West Iran

Ian A. Brookes

ROM
ROYAL ONTARIO MUSEUM

Royal Ontario Museum
Publications in Archaeology

The Royal Ontario Museum publishes two series in the field of archaeology—Monographs, a numbered series of original publications, and Papers, a numbered series of primarily shorter original publications—as well as unnumbered major monographs. All manuscripts considered for publication are subject to the editorial policies of the Royal Ontario Museum, and to review by persons outside the Museum staff who are authorities in the particular field involved.

Ian A. Brookes is associate professor, Department of Geography, York University, North York, Ontario, and research associate, West Asian Department, Royal Ontario Museum.

Cover: Spring on the Mahidasht. Distant limestone ridge at 2300 m fronted by coalescent, cultivated alluvial fans which sweep down to lowland with pastures at 1400 m. Transhumant herders gathered at black tent on outskirts of village (far right) sprawled over low *tepe*. 17 May 1978.

Canadian Cataloguing in Publication Data
Brookes, Ian A.
The physical geography, geomorphology, and late Quaternary history of the Mahidasht Project area, Qara Su Basin, central west Iran

(ROM Mahidasht project ; v. 1)
Includes bibliographical references.
ISBN 0-88854-346-8

1. Geology, Stratigraphic — Recent.
2. Archaeological geology — Iran — Mahidasht Region.
3. Physical geography — Iran — Mahidasht Region.
I. Royal Ontario Museum. II. Title. III. Series.

QE696.B76 1989 551.7'93'095552 C89-093253-0

100 Queen's Park, Toronto, Canada, M5S 2C6
ISBN 0-88854-346-8
PRINTED AND BOUND IN CANADA AT THE UNIVERSITY OF TORONTO PRESS

Contents

Illustrations

Figures

Plates

Tables

Acknowledgements

I owe an enormous debt of gratitude to Louis D. Levine, associate director, exhibits, Royal Ontario Museum, director of the Mahidasht Project, who invited my participation in its first two field seasons—not only for the invitation, but for making that participation so rewarding, scientifically and personally.

The work was financed by generous grants from the Canada Council, (now Social Sciences and Humanities Research Council of Canada), supplemented by funds from the Royal Ontario Museum, Toronto.

The Iranian Centre for Archaeological Research, Tehran, generously assisted in matters relating to the project's support and conduct.

Three radiocarbon dates were kindly provided by P. LaSalle, Service d'Exploration géologique, Ministère des Richesses naturelles, Québec. One thermoluminescence date was kindly provided by the Archaeometry Laboratory, Oxford University.

Robert A. Finnie, formerly of the West Asian Department, Royal Ontario Museum, competently assisted me in the field in 1978, subsequently ran many analyses at York University, Toronto, and has been helpful in many ways since. The Museum's Exhibit Design Services reproduced my photographs, while the staff of Publication Services and my editor, Andrea Gallagher Ellis, have nursed the manuscript, not to mention its author, to final publication.

York University has provided me with many of the essentials for completion of this work—a salary of course, office and laboratory, competent secretaries, and especially cartographers. I thank the institution and all these individuals. Two reviewers made comments which improved the report, but remaining errors and flaws are entirely my responsibility. The Royal Ontario Museum's West Asian Department has always been a good-humoured sanctuary from other pressures, bolstering my sometimes faltering belief in the existence of a place called "The Mahidasht" so many (too many) years after the expedition reluctantly left it.

Abstract

This report deals with several topics fundamental to the interpretation of archaeological information on past human settlement patterns, population densities, and adaptive strategies in the largest intermontane basin in the Zagros Mountains of Iran, a 4900-km^2 area centred on 34°30′N, 46°30′E.

The report first synthesizes the author's field data and previously published information on the modern environment to provide a base for comparison with past conditions. It then defines and characterizes landform regions with respect to morphology, composition, and origin. It goes on to describe and interpret a sequence of Quaternary sediments, soils, and erosion surfaces, and from this infers a sequence of environmental changes. Finally, it relates these changes to data on the size and distribution of archaeological sites dating from ca. 5000 B.C. to late medieval times. The report concludes that geological processes have played a more important part than cultural ones in explaining them.

Environments of the Holocene are seen as relatively passive influences on human occupation and land use, in a lowland setting dominated first by alluvial aggradation of muds, then by pedogenesis on a stable land surface. This setting was catastrophically disrupted roughly a thousand years ago by a basin-wide flood which eroded occupation sites from piedmont zones. Subsequent lowland accumulation of up to 10 m of muddy alluvium has buried sites which were abandoned before, or as a result of, the flood, and which were lower in height than younger alluvium. Alluviation has also buried those which, although occupied, failed to grow in height faster than recent alluvial aggradation. The distribution of sites with, for example, pre–Bronze Age material (older than ca. 3200 B.C.) as a sole, major, or minor component is used to evaluate the hypothesis of alluvial burial of sites.

Introduction

Objectives

This report first provides a background to the modern natural environment of the Mahidasht Project area through a consideration of its geology, climate, hydrography and hydrology, soils, and vegetation, and proceeds to an interpretation of its later geological history, focusing on the Late Pleistocene and Holocene epochs. The work was undertaken to provide a palaeoenvironmental sequence potentially to be matched with an archaeological record of cultural change in the Mahidasht. Its conclusions, however, have so far added more to the interpretation of archaeological site size and site distribution than to explanations of the cultural sequence. This partly reflects the present incompleteness of archaeological data reduction and interpretation, but at time of writing it appears that the palaeoenvironmental sequence will remain as its main contribution.

Fieldwork for the study was done from 19 July to 15 August 1975 and from 12 May to 15 July 1978, a total of 93 days. In the first period reconnaissance work established the outlines of the geomorphology and alluvial stratigraphy. The second focused on detailed mapping, testing of the stratigraphic inferences, studies of archaeological site morphology and stratigraphy, and brief visits to Kangavar and Holailan valleys for comparative studies. With the areal pattern of landforms and surface materials established and the stratigraphic sequence reasonably secure, the close of the 1978 season left the canvas begging for completion with detailed work on the many questions raised. Political events have prevented our return, so that, like others to follow, this report should be seen only as full and secure in its evidence and interpretations as the limited fieldwork has allowed.

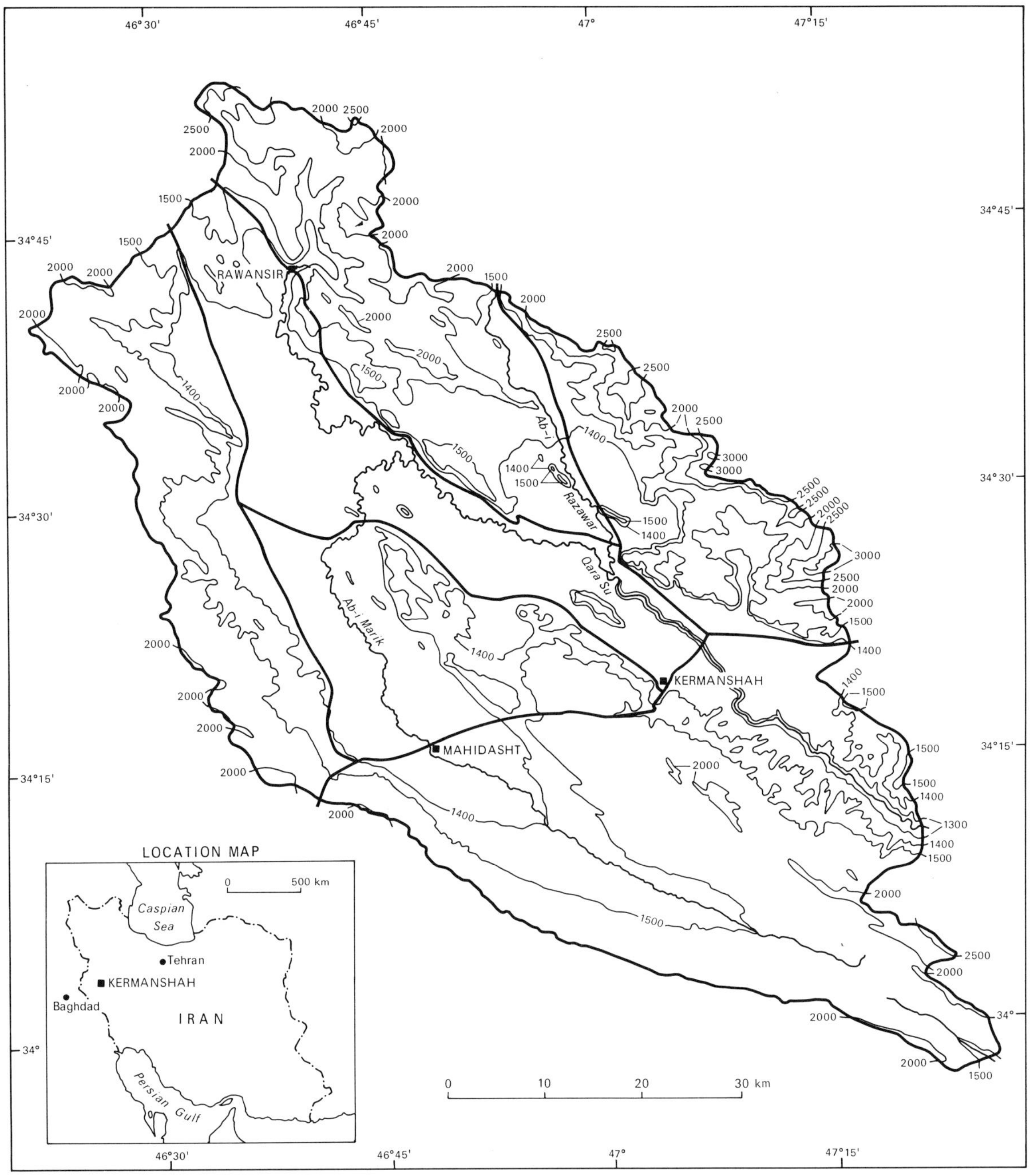

FIG. 1. Mahidasht Project area, Iran, showing major settlements, major roads, perennial streams, contours at 1300 m, 1400 m, 1500 m, 2000 m, 2500 m, and 3000 m. Inset shows location in Iran.

Location

The Mahidasht Project area lies in central west Iran, around the town of Kermanshah (now Bakhtaran; 34°17′N, 47°1′E, Fig. 1). The name "Mahidasht" is more strictly applied to a large village on the Shahabad–Kermanshah road where it crosses the Ab-i Marik, and less formally is given to the alluvial lowland of that river. For project purposes, the name is applied even less strictly to the study area which encompasses that lowland, as well as two contiguous ones to the north and east, together with the intervening and surrounding hills and mountains. The boundary of the project area was drawn along the drainage divide of the Qara Su above its confluence with Gamas-i Ab. An area of 4900 km^2 ($\pm$1%) was determined by accumulating the areas occupied by seven landform types within 5 km $\times$ 5 km UTM grid squares. These areas in turn were determined using a random dot overlay. The watershed approach to boundary definition was adopted because of the functional integrity of watersheds from an ecological standpoint and potentially from a cultural one also. The upper basin of Ab-i Razawar above Kulichan was however excluded because its only link with that of the Qara Su is by a narrow gorge and likely has not been strongly linked in an economic sense. Conversely, a small area above the source of the Ab-i Marik was included since links there have most likely been with settlements in this basin. Administratively, the area lies wholly within the *ostan* ("province") of Kermanshahan (now Bakhtaran), of which Kermanshah is the largest urban place and the capital with an estimated population (1976) of 290 861 (Encyclopedia Britannica, 15th ed., s.v. "Bakhtaran").

Communications

Kermanshah is regularly serviced by Iran Air flights and express bus service from Tehran. The main Baghdad–Tehran highway crosses the area through Kermanshah. Other paved roads link it with Sanandaj and Paveh to the northwest; Bisitun and Kangavar to the east; and Shahabad, the Persian Gulf, and Iraq to the west. A newly aligned road, in 1978 not yet paved, serves the western flank of the area from the main trans-valley road to the northwest edge, and a spur from Kermanshah aligned west and northwest joins it in the northwest of the lowland. The area is densely reticulated with one-lane, dry-weather tracks linking the dense network of 485 villages. In parts of the lowland these tracks may be only temporary, as field boundaries are sometimes moved. As well, the tracks may cross irrigation ditches, not always clearly visible and therefore hazardous to vehicles. The only impediment to efficient planning of geological fieldwork proved to be the paucity of river crossings, although the attentions of village dogs also meant that some localities were visited rather too briefly.

Toponymy

Place-name usage in this report adheres to the English transliteration of Kurdish names printed on U.S. Army Map Service, 1:250 000-scale map sheets (N138-7,8,12; 1958). Diacritical marks are, however, omitted.

Research Methods

Geomorphological studies have included airphoto interpretation, checking of landform boundaries and logging of stratigraphic sections in the field, and laboratory work. Airphotos at 1:21 600 scale were made available for the entire area through the Iranian Centre for Archaeological Research, Tehran. Geomorphic and cultural features were interpreted on photo overlays, using a $\times$ 2 pocket stereoscope. This information was transferred to 1:50 000-scale maps with 20-m contour interval, made available through the Iranian National Cartographic Centre. These maps were specially produced for irrigation and land-use planning and show perennial and intermittent drainage, spot heights, roads and tracks, settlements, major irrigation channels, and extent of cultivation. Lack of this map coverage in the far northwest, north of Rawansir, and in the far west, north of the Qara Su–Gamas-i Ab confluence, was made up by attaching enlargements of the required areas from the U.S. Army Map Service map sheets (1:250 000 scale).

Landform boundaries interpreted from airphotos were checked and adjusted in the field, and surface characters noted. Sections identified on photos were visited, logged, and sampled. Some morphological and stratigraphic work was done on archaeological sites also.

Laboratory work included analysis of particle size, organic matter, calcium carbonate, and clay mineralogy. Radiocarbon dating was done externally.

Physical and Human Geography

Generalized Tectonic History

The project area lies astride a major boundary between tectonic zones of the Zagros Mountains, which are a segment of the trans-Eurasian Alpine Orogenic System. The tectonic evolution of this system has been outlined by Dewey et al. (1973) and a short synthesis for the Iranian region is given by Takin (1972).

An earlier orogenic cycle, the Hercynian, had ended about 220 million years (Ma) ago with the closing of an ocean by northward movement of one or more continental blocks now in central Iran. This movement, while closing one orogenic cycle, opened another, the Alpine, as an ocean widened between Iranian crustal blocks and the Arabian block. This was part of the Tethyan Ocean, which widened as the northern supercontinent, "Laurasia" (later to separate as North America and western Eurasia), and the southern "Gondwanaland" (later to separate as the southern continents) drifted apart. This drift phase lasted about 40 Ma and Tethys is believed to have widened to about 4000 km in this time (at an average rate of 10 cm per year).

Between 220 Ma and 65 Ma ago this ocean was closed as the Afro-Arabian and Indian continental blocks drifted north and intervening oceanic crust was subducted beneath the Eurasian block. During this closure the advancing margin of the Arabian continent was a shallow, subsiding shelf on which thick, shallow-water, marine sediments were being deposited in linear, fault-bounded basins. The opposing Iranian side of the closing ocean consisted of a mosaic of small continental blocks broken during the initial rifting and drift from Arabia. Between these blocks sedimentation was expectably complex.

Final closure of the High Zagros Alpine Ocean (this section of Tethys) occurred between 75 Ma and 65 Ma ago, when igneous oceanic crust and mantle rocks (ophiolites) as well as deep-sea radiolarian cherts were thrust southwest in thick slices. The ophiolites now appear east of the High Zagros in the Rezaiye-Esfandagheh Orogenic Belt while the cherts are found east of the Main Zagros Thrust. In central Iran, the micro-continents were consolidated by compression from the southwest against a rigid northern continental block.

Between 65 Ma and 20 Ma ago, with the true ocean basin of Tethys now closed in the Zagros region, and tectonically thickened crust elevated along its axis, denudation by southwest-flowing streams fed rapid deposition of "flysch" sediments to a shallow foreland trough, bounded to the east by the Main Zagros Thrust. Postclimactic basaltic volcanism east of the thrust between 47 Ma and 42 Ma ago reflected relative crustal stretching of the new orogenic "pile" as northeast-directed Arabian plate movement slowed.

The most recent 20 Ma have witnessed simple folding of the Arabian continental shelf sequence of limestones and shales, initiated upon opening of the Red Sea–Gulf of Aden, and a counter-clockwise rotation of the independent Arabian block. This has been accompanied by continuing denudation of the earlier-deformed zones as well as contemporaneous denudation of the folded rocks to produce the modern mountain and basin landscape.

Geology of the Study Area

The study area embraces three large-scale tectonic zones of the Zagros Orogen (Fig. 2 and Pl. 1). East to west these are, the Rezaiye-Esfandagheh Orogenic Belt (REOB), the Zagros Crush Zone (Coloured Mélange and Radiolarites) and the Zagros Folded Belt (Stocklin and Nabavi, 1972; Takin, 1972).

The REOB is bounded to the southwest by the Main Zagros Thrust, which dips northeast at a steep angle, while the other two zones are separated by a minor thrust, which strikes northwest through the area but occasionally dies out along strike.

The *REOB* in the study area is divided into two subzones. An outer (southwest) zone consists of southwest-directed, massive thrust slices of Middle Cretaceous crystalline limestone with minor chert and shale (Km). Any original bedding has been obliterated by deformation, but within each thrust slice curvilinear ridges impart a smaller-scale physiographic grain to an otherwise undiversified mountainous terrain. The inner (northeast) subzone comprises Eocene (47–42 Ma ago) gabbro and basalt (Es β) with Miocene postclimactic molasse sediments (m).

The *Zagros Crush Zone* is divided into three subzones: (*a*) a southwestern zone of thrust-bounded slices of Late Cretaceous limestone (Km-l); (*b*) a northeastern zone of infolded masses of Coloured Mélange, consisting of ophiolites and radiolarian cherts (Ra), tectonically emplaced during Late Cretaceous thrusting of

REOB; (*c*) a middle zone of radiolarian cherts and detrital limestones (Ra).

The *Zagros Folded Belt* consists of a Jurassic to Miocene sequence of shallow shelf sediments, mainly carbonates (J, Gu, Tal) with significant intervening formations that represent deposition of clastic sediments eroded from the newly thrust carbonates and ophiolites of REOB (Am, Ka). These latter formations are of Late Cretaceous and Eocene age. This entire sequence was folded during the Zagros Orogeny in Late Tertiary time, while contemporaneous denudation produced strong northwest-trending ridges and interconnected basins.

Climate

An up-to-date review of the main features of the climate of the Near East appears in Takahashi and Arakawa (1981), while the Iranian area is treated in Ganji (1955, 1965, 1968). Broadly, the Iranian climate is influenced by three factors: (*a*) a mid-latitude position, astride the subtropical zone, experiencing polar and tropical air-mass flows seasonally; (*b*) a continental location which increases the annual range of temperature and attenuates precipitation derived from oceanic sources; (*c*) generally high elevation, with a mountainous periphery which intercepts westerly and northerly airflows, producing a pronounced precipitation "shadow" in the interior, and which lowers temperatures while enhancing precipitation to windward.

Winter atmospheric pressure distribution is dominated by the Asiatic Anticyclone, with mean "lows" over the southern Caspian Sea and southern Iran, which produces a mean northerly flow in the east, an easterly in the south, and an east-northeasterly in the west. These flows, based on January pressure distribution, are frequently overridden by the passage of lows steered over Iran by the subtropical jetstream. They give the winter climate its distinctive cool, moist regime.

In summer pressure is dominated by intensified lows over southern and southeastern Iran, a weak low over central west Iran, and a weak high over the Caspian Sea. This pattern induces a dominant southeasterly flow over central and east Iran, a southerly over the south and southeast, a southwest and northwest over the west.

The region of western Iran encompassing the study area experiences winter (January and February) mean temperatures of 2°C–3°C, due to low insolation, polar air-mass dominance, and elevations of 1000 m–1500 m in intermontane valleys (2000 m–3000 m on summits). Summer (July and August) mean temperatures range from 25°C to 30°C in valleys.

Precipitation occurs only in winter (92% October to May, 75% November to April), and is brought by depressions originating over the Mediterranean Sea, sometimes regenerating over the northern Persian Gulf. Upper westerly "shallow-waves" bring low-intensity precipitation in mid-winter, whereas cold troughs cross the Zagros more slowly and bring heavier precipitation in late winter and spring (Brichambaut and Wallén, 1962).

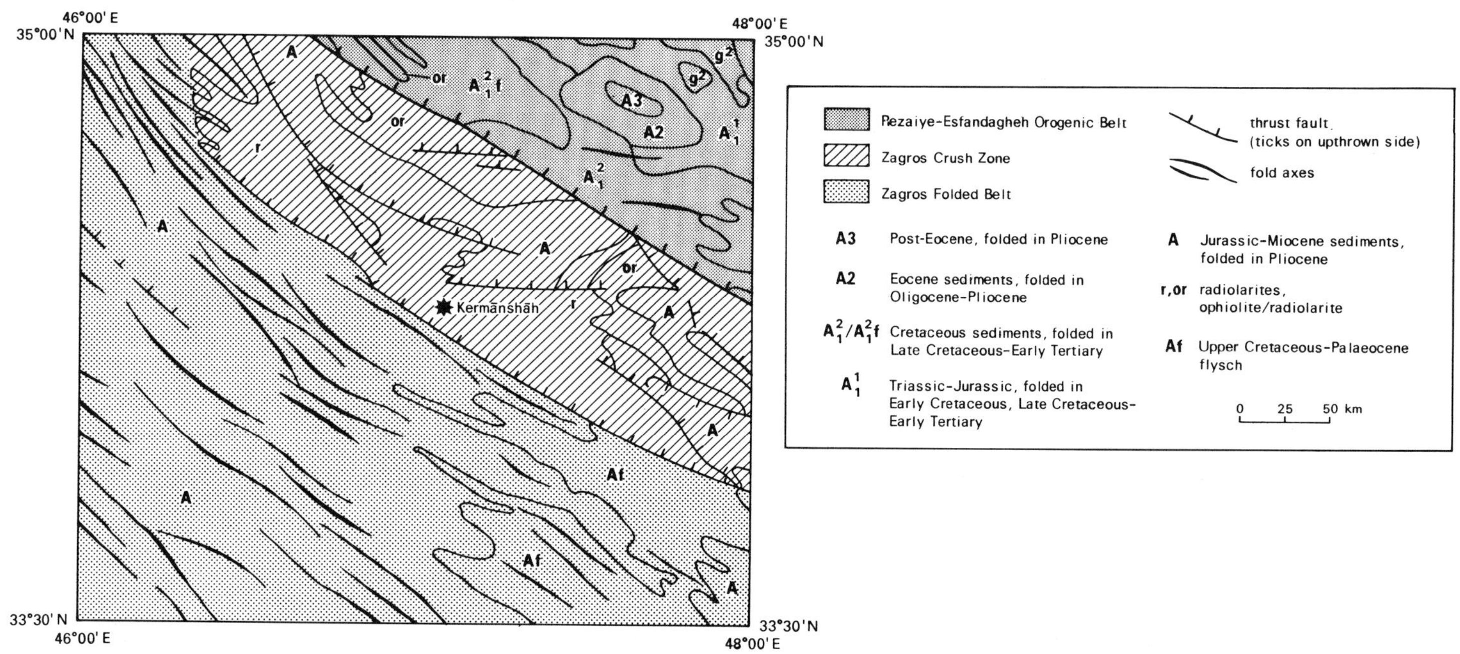

FIG. 2. Tectonic zones of the Zagros Orogen in the Kermanshah region (after Stocklin and Nabavi, 1972).

TABLE 1. Mean Monthly Temperature and Precipitation at Kermanshah.

Month	Temperature (°C) 1941–1960*	Precipitation (mm) 1951–1960*	Temperature (°C) 1956–1972†	Precipitation (mm) 1956–1972†
Jan.	2.5	37.8	1.8	65.6
Feb.	2.2	46.3	3.2	57.0
Mar.	6.6	67.8	7.9	78.9
Apr.	11.0	68.0	11.9	98.4
May	16.5	30.0	16.8	44.2
June	22.0	2.9	22.0	2.0
July	26.2	0	26.8	0
Aug.	25.8	0.1	25.9	0.6
Sept.	21.0	1.3	21.3	1.3
Oct.	16.5	12.3	15.6	27.0
Nov.	8.8	55.3	9.0	53.7
Dec.	2.7	50.9	3.9	50.8
Total		372.7		479.5
Mean	13.5		13.8	

*SOURCE: Ganji, 1968.
†SOURCE: Iran Ministry of Water and Power, 1956–1972.

Temperature and precipitation data for Kermanshah (at 1322 m, near the lowest and most easterly point in the study area) are given in Table 1. Two sets are given for comparison of earlier and later periods, but only those for 1956–1972 are shown graphically (Fig. 3). The comparison shows that winters have been warmer and wetter in the later period, and that, while the mean annual temperature has not changed, precipitation has increased by 28%. Yearly data show that winters were drier than the 1956–1972 mean up to 1965, but were wetter after that. This may parallel a tendency to increased storminess in European mid-latitudes since the mid-1960s.

This same data set shows that mean temperatures are more variable in winter months than in summer, with January means occasionally below freezing. From 1957 to 1970 frost-free days ranged from 245 to 283, with a mean of 266.7 (Table 2). Even in warm winters, then, 80 days experience frost near the lowest elevation of the area, so that higher zones likely experience considerable freeze-thaw activity, of some significance for rock weathering and soil structure.

Precipitation data are available for only one other station in the study area—Rawansir, in the northwest—so that no reliable isohyets can be drawn. Unpublished short-term records for several other stations were perused at government offices in Kermanshah, but they showed no significant differences that could be explained by topographic "shadowing" or depression-track variation.

Daily precipitation data for Kermanshah (1956–1972) are plotted on a probability scale in Figure 4. This shows that the median value is 4 mm, and that only 10% of precipitation days (an annual average of 4.4 days) experienced more than 24 mm. The trend of precipitation intensity from 1957 to 1970 is shown in Figure 5. Notably, years of greater-than-average intensity do not coincide with years of above-average annual runoff or annual flood peaks on the Qara Su (see next section).

Snowfall days (1957–1970) are shown in Table 2, and clearly vary considerably from year to year. Data for Rawansir, in the northwest of the area but only 200 m higher than Kermanshah, show an erratic pattern of more and fewer snow days, but it could be expected that many more would occur in the mountains. The streamflow record, discussed in more detail in the next section, does not, however, bear this out. Winter daily precipitation at Kermanshah (which, being mostly frontal, can be expected to be basin-wide on most days) is, in nearly every case, followed three to four days later by a peak of streamflow at the gauge nearby. Since this interval is theoretically required for channel flow from the extremities of the Qara Su basin (at an estimated rate of 0.3 ms^{-1}) to reach the gauge, it appears that a large majority of precipitation events, even in the mountains, yield rain rather than snow. Further, there is no spring snowmelt runoff peak on the Qara Su. The baseflow curve steadily rises through the winter; precipitation events are reflected as delayed streamflow "spikes" superimposed on it. This again suggests not only rainfall rather than snow, but also that what snowfall there is probably melts slowly to percolate into the limestone bedrock, thus contributing to winter baseflow increases. This is borne out by the response of the large spring at Rawansir to precipitation (Fig. 6).

TABLE 2. Frost-free Days and Days with Snow or Sleet, Kermanshah (elevation 1322 m).

Year	Frost-free Days	Days with Snow or Sleet
1957	283	28
1958	271	5
1959	245	13
1960	247	2
1961	250	6
1962	274	4
1963	277	5
1964	265	12
1965	270	12
1966	278	3
1967	253	13
1968	275	16
1969	278	9
1970	271	2
Mean	266.7	9.3

SOURCE: Iran Ministry of Water and Power, 1956–1972.

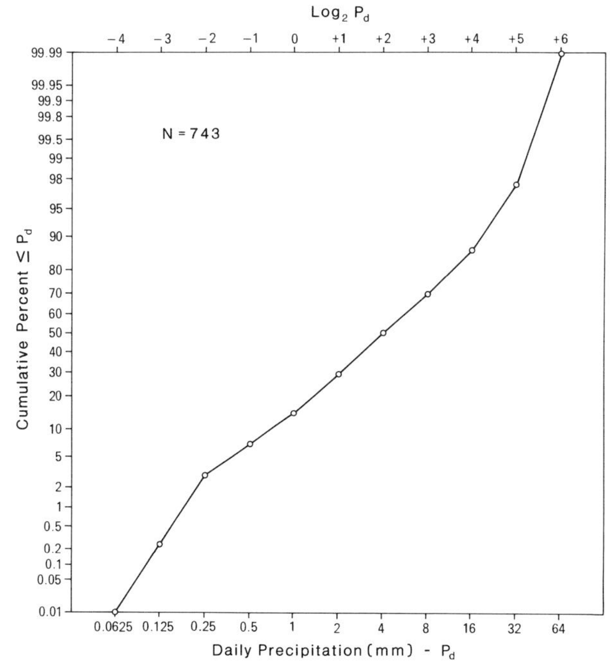

FIG. 4. Probability plot of daily precipitation (mm), Kermanshah (1956–1972; N = 743 precipitation days).

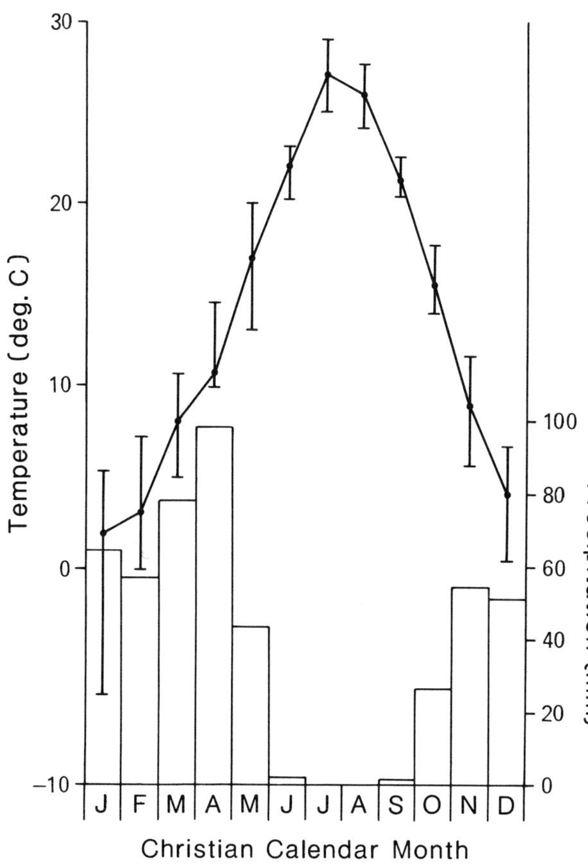

FIG. 3. Monthly mean, maximum, and minimum temperatures, and precipitation, Kermanshah (1956–1972).

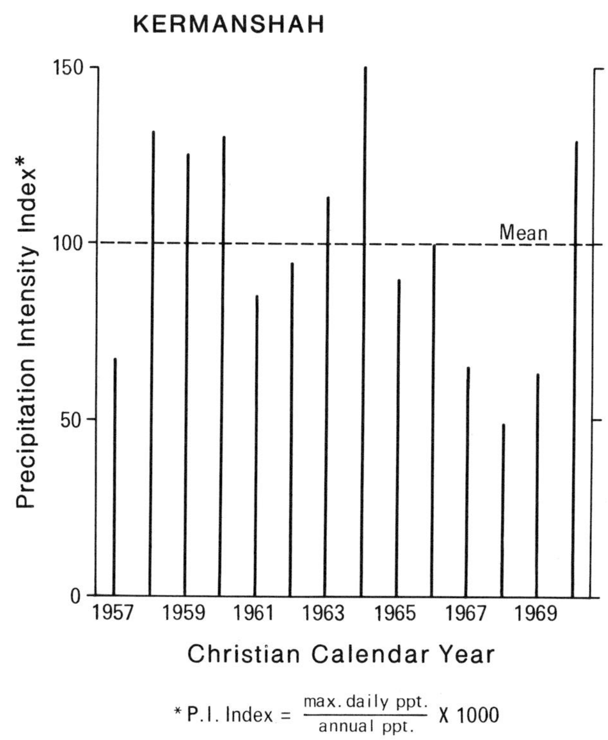

FIG. 5. Precipitation intensity index, Kermanshah (1957–1970).

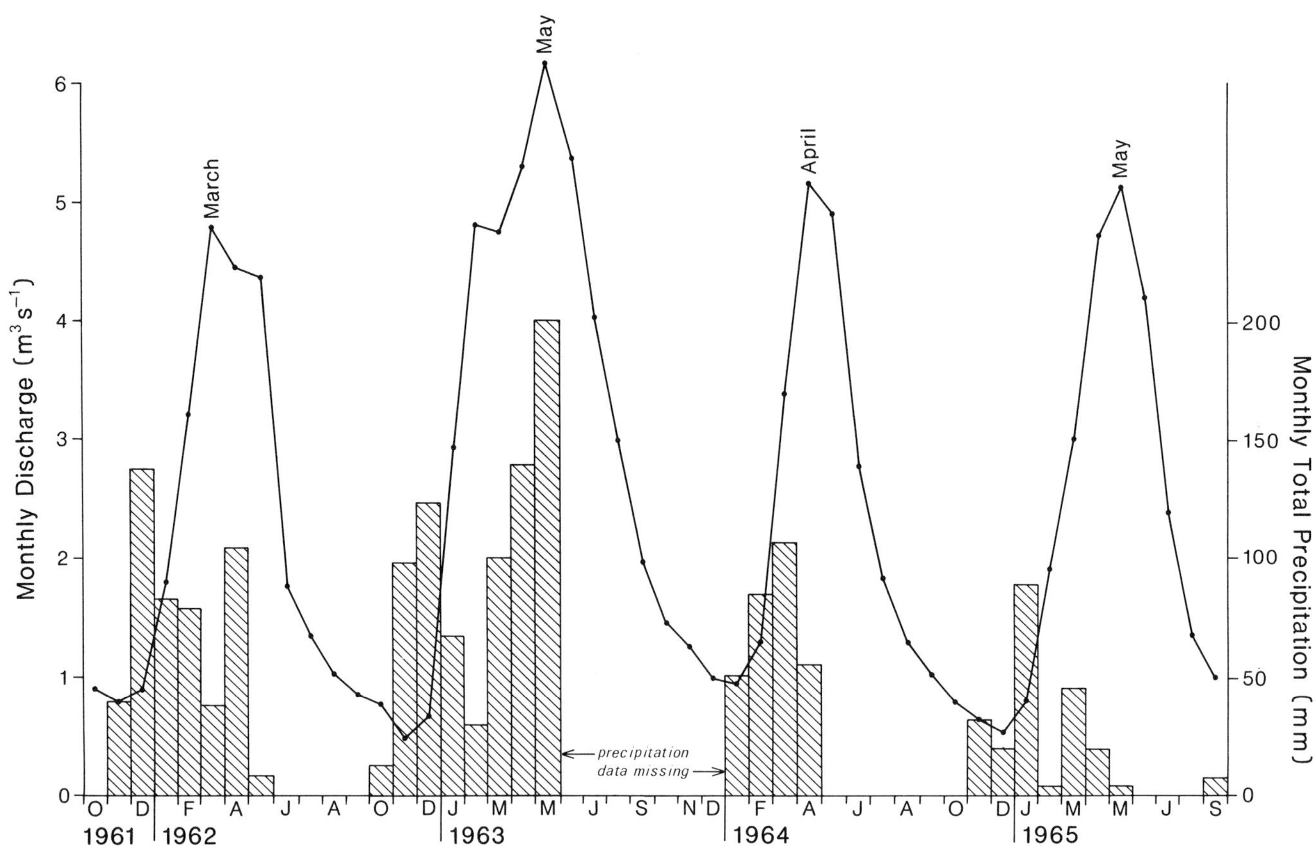

FIG. 6. Monthly total discharge (m^3s^{-1}) of Rawansir spring 1961–62 to 1965–66 water years, and monthly total precipitation at Rawansir.

Hydrography and Hydrology

With a mean annual precipitation at Kermanshah, near the lowest point in the area, of 373 mm (1951–1960)—or 480 mm (1956–1972; see Table 1)—the study area is certainly well watered, although the prevalence of permeable carbonate bedrock and extensive hilly and mountainous terrain means that water tables are low, and only trunk streams are perennial (Pl. 2). The dense network of intermittent streams experiences flow only from March to May when the water table is at its highest and spring rains cause high runoff rates.

Bedrock lithology and structure are primary controls on the drainage pattern. There are roughly 250 perennial springs in the area, an average of one every 20 km^2. They are distributed in three well-marked zones (Pl. 2). One lies along the contact between Cretaceous limestones of the Zagros Crush Zone and colluvium/alluvium on the northeast. This zone includes the large Taq-i Bostan, Lalabad, and Rawansir springs, and those along the eastern edge of the Ab-i Razawar lowland. Another zone contains more uniformly distributed springs over the outcrop of the Coloured Mélange and Radiolarites. It comprises two distinct subzones, one to the northwest and southeast of Kermanshah, and one in the far northwest, west of Rawansir. A third zone of springs is found within the outcrop of the Amira Formation. This consists of clastic flysch sediments and outcrops along the southwest flank of the study area, below the boundary ridge. In this zone springs on the west side of the Ab-i Marik valley occur where the Amira Formation dips upslope. Groundwater percolation is therefore down-dip into sandy aquifers. Springs arise where intermittent stream beds intersect the water table.

Drainage is either non-existent or of low density in three zones (Pl. 2). One coincides with the outcrop of Cretaceous limestones of the Zagros Crush Zone in the northeast. Here, drainage is entirely subterranean, since the rock is intensely karstified (Waltham and Ede, 1973; Maire, 1978). Waters emerge at the contact with unconsolidated deposits, as at Taq-i Bostan. Another "dry" zone coincides with the Cretaceous limestones beneath Kuh-i Sefid and its northwest continuation. In the Sefid the rocks dip northeast at 15°–20°, so that groundwater passes along the dip to emerge at the thrust-faulted contact with the Coloured Mélange, or at the contact with alluvium, as at Sarab-i Nilufar. A third zone of low drainage density is the extensive alluvial lowland outside the spring basins. Here, geological discontinuities are absent or buried and water-table slope is too low to maintain a hydraulic head. A large spring basin is, however, found in the northern Marik lowland west of the river. Twenty springs are mapped; nine of these are aligned, suggesting a buried geological contact.

The study area is drained by only four perennial streams, the Qara Su, the Ab-i Marik, the Ab-i Razawar, and the Ab-i Khan (Pl. 2). Only the Qara Su is gauged, at three places: Doab, near the Ab-i Marik confluence; Pol-i Khoneh near the Ab-i Razawar confluence; and Gharbaghestan, near the Gamas-i Ab confluence (Pl. 2). Hydrographs of daily discharge at Gharbaghestan and superimposed graphs of daily precipitation at nearby Kermanshah for the water years 1958–59 to 1969–70 are included (see Appendix). Those for the years with lowest (1961–62) and highest (1968–69) flood peaks are reproduced as Figure 7. Annual runoff, monthly discharge, and annual discharge at this downstream station are shown in Figures 8, 9, and 10.

The hydrographs show that most discharge peaks follow high precipitation events by three to four days. If a mean flow velocity is estimated at 0.3 ms^{-1}, this is the time required for stream water to travel from the

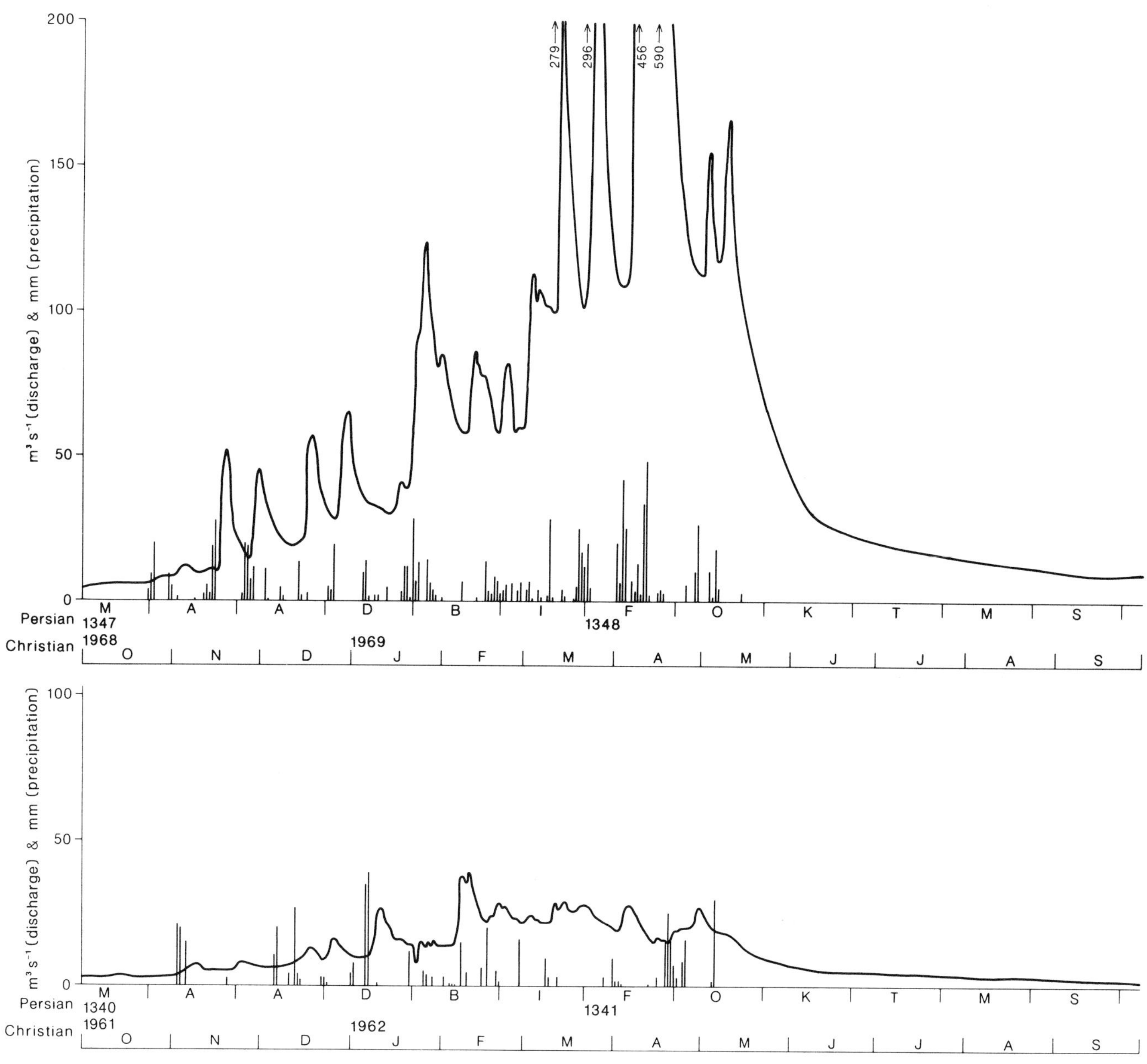

FIG. 7. Hydrographs of daily discharge (m^3s^{-1}), Qara Su at Gharbaghestan, and bar graphs of daily precipitation (mm), Kermanshah, for a low flood-peak year (1961–62) and a high flood-peak year (1968–69). See also Appendix.

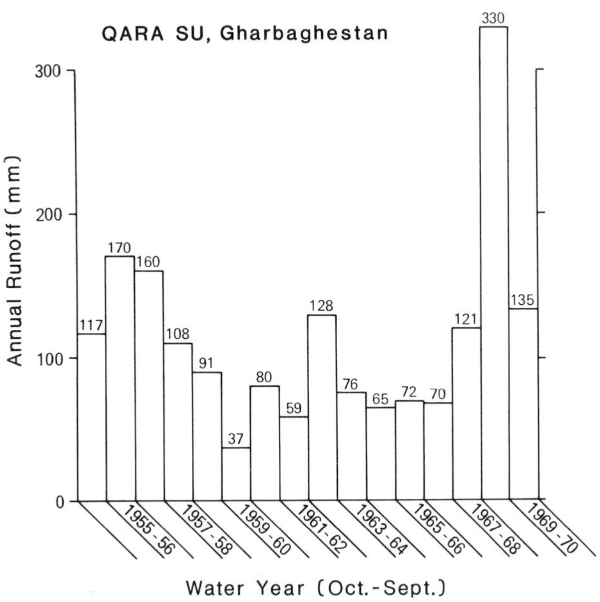

FIG. 8. Annual runoff (mm), Qara Su at Gharbaghestan, 1954–55 to 1969–70 water years.

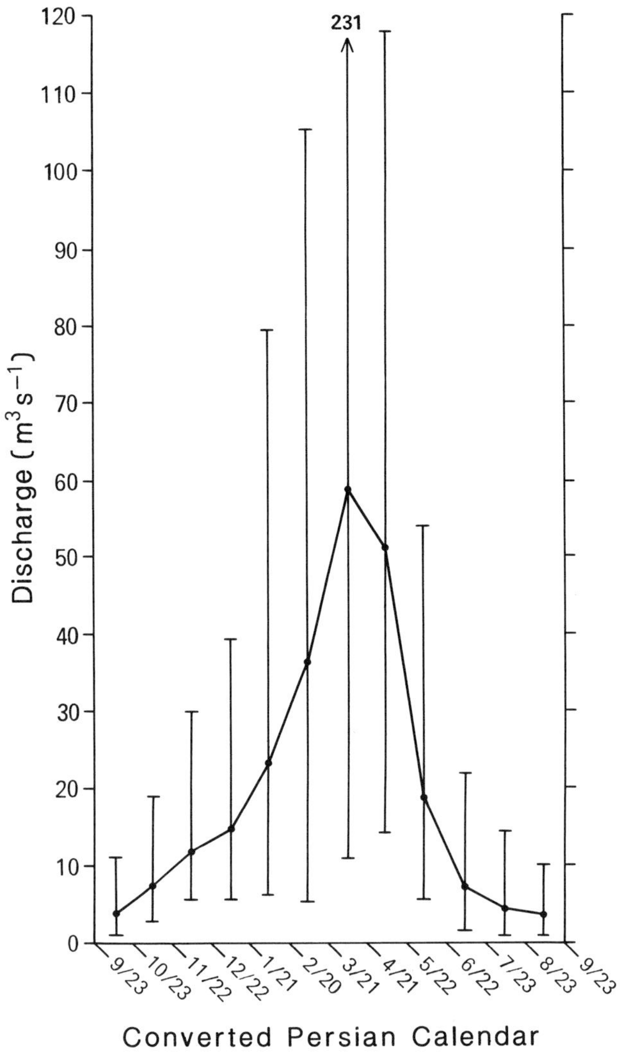

FIG. 9. Mean, maximum, and minimum monthly discharge (m^3s^{-1}) of Qara Su at Gharbaghestan, 1954–55 to 1969–70 water years.

extremities of the Qara Su, Marik, and Razawar basins to the gauge at Gharbaghestan. Therefore, the discharge peaks are interpreted as responses to basin-wide precipitation, as expected if this was yielded from passing depressions. Expectably, early winter rains do not produce high discharge peaks because most of the precipitation contributes to soil-water and groundwater recharge. As groundwater levels rise (indicated by the rising baseflow component of discharge shown by the hydrographs in the Appendix, and by the response of Rawansir spring shown in Fig. 6), discharge peaks more closely reflect precipitation events. Peaks lower than predicted by precipitation levels suggest that precipitation was not basin-wide in those events. Those higher than expected probably indicate that precipitation at Kermanshah was less than that over higher and more distant areas of the basin.

As discussed in the climate section, the fact that throughout the winter discharge peaks follow precipitation events by three to four days indicates that much of that precipitation must be occurring as rain, even though Kermanshah and Rawansir data show that snowfalls are common there (and even commoner at higher elevations). There are, however, some December and January precipitation events that produce no response in streamflow; so they may be largely snowfalls.

The 12-year period of discharge record available here (see Appendix) hardly permits statistical analysis of flood peaks, but fortunately includes the exceptional flood of April 1969. This followed an exceptionally wet winter, abnormal not in length or number of precipitation days, but in that each precipitation day experienced high amounts. Baseflow had thus risen higher than many years' peak flow, and daily precipitation ranging from 40 mm to 50 mm was superimposed. During the succeeding five-month drought, baseflow did not fall below 10 m^3s^{-1}, two or three times its value in other years, so that late-1969 and early-1970 discharge peaks were higher than precipitation would have predicted. Geomorphic effects of the annual flood are discussed under the heading Alluvial Plain.

Soils

Maps of the soil series of the Ab-i Marik watershed at 1:50 000 scale were consulted at the Kermanshah office of the Iran Soil Survey. Extrapolation of series descriptions to homologous geomorphic zones in other parts

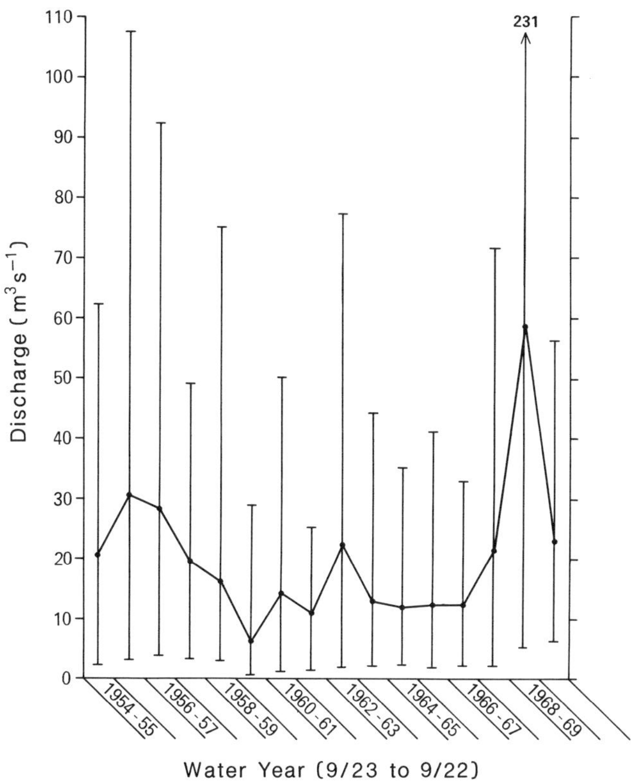

FIG. 10. Annual mean, maximum, and minimum discharge (m^{3}s^{-1}) of Qara Su at Gharbaghestan, 1954–55 to 1969–70 water years.

of the study area provides a reasonably accurate and complete picture of its soils. The primary criterion of mapping here is geomorphic, since both topography and parent materials are subsumed in it, as are secondary characters of drainage, stoniness, and erodibility. Climate is not an important variable in explaining soil differences in such a small area. Also, vegetation long ago lost its natural aspect here under the combined influences of axe, plough, and grazing animals (Zohary, 1973). Age of parent materials varies considerably, so it would be expected to influence soil types. However, the older the parent material, the more complex the sequence of geomorphic events that have subsequently affected it, so that simple histories are not to be expected. Such complex soils are also likely to be either buried or deeply truncated.

Mention of geomorphic history brings into focus an influence which, second to parent material, dominates soil characteristics. The surface alluvial soils of the area are developed on materials which have been deposited only in the last thousand years, and the gravelly alluvial soils of the piedmont are often only thinly developed over parent materials which were severely eroded to yield the young alluvium of the lowlands. Some piedmont soils on colluvium, however, are undisturbed and deep.

Mountain and Hill Soils

The dissected, steeply sloping, mountainous terrain marking the northeastern rim of the Qara Su basin is almost devoid of soils. Where gentle slopes or depressions have accumulated the eroded products of weathering and aeolian influx, soils are found only in small pockets between boulders and rocky ribs. The soil pockets consist of angular carbonate pebbles in a silty loam matrix which retains little moisture. Development is limited to an incipient A-horizon over a structureless C-horizon.

Over smoother-topped uplands in the central and northwestern areas, less destructive geomorphic processes have allowed a more continuous, but still shallow and stony, soil to develop. By comparison with other uplands, these areas are extensively mantled with about a metre of reddish, stony loam, known as *terra rossa.* Hillslope erosion however has often stripped this loam and transported it to valley floors. In parts of the northwestern area, the density of valleys is low because of the permeability of the bedrock, so that broader interfluves have preserved stony *terra rossa.*

Hills of the southwestern flank of the study area are geologically more variable. Frequent geological contacts produce a dense pattern of springs which feed a similarly dense drainage network. Interfluves are thus small, and steeper valley sides more extensive, so that soils are eroded, stony, and skeletal.

Piedmont Soils

The piedmont between the mountain and hill zones and the lowland zone has a complex geomorphic history, variable in space and time. Its landform elements are thus distributed in an intricate mosaic.

Pediments and alluvial fans are the dominant landforms in this zone; their genesis will be discussed in the geomorphology section. Soils on these landforms are classified as the Rawansir Series. The parent material is a loamy, calcareous gravel, while topography is planar, sloping towards the lowlands at 2°–5°. Soils are usually less than a metre deep, uniformly reddish-brown in colour, and cloddy in structure with a high percentage of stones, which are commonly cleared.

Valleys dissecting the pediments and fans are the sites of the Kunuj Series, developed on reddish, stone-free loams, believed to have been eroded from soils earlier developed on pediment interfluves. These soils have *p*H values of 7.7 to 7.9, similar to the Rawansir Series, but their carbonate content is lower; they are well drained and highly arable. Access to irrigation water is the only limitation to their extensive cultivation.

At the lower ends of piedmont valleys Kunuj soils give way to the Tiram Series, developed on reddish-brown silty clay loams over clays. Parent materials represent the finer-grained sediments flushed from higher in the valleys and are indistinguishable texturally from alluvial soils in the lowlands.

Lowland Soils

Lowlands are up to several kilometres wide and slope imperceptibly from the piedmont to trunk streams and down-valley. The Marik lowland, for example, slopes at 1:900 longitudinally and 1:600 transversely.

The major distinction amongst lowland soils is their derivation from riverine alluvium or from alluvium deposited in spring basins and reworked by spring overflow channels. The latter belong to the Burbur Series. These are extensive around Malekshah in the northwest alluvial area, around Lalabad north-northwest of Kermanshah, and around Sarab-i Nilufar, west-northwest of Kermanshah. In these basins and outflow zones, alluvial aggradation by the trunk streams has impeded free drainage from the springs, so that overflow channels anastomose across nearly flat topography, itself the result of sediment build-up in the basins. The Burbur soils are grey silty clays over clays with *p*H values of 7.5 to 7.7. They show the highest carbonate levels of any series in the area, which reflect alkali efflorescence from slow-draining topsoils and precipitation from standing water.

Riverine alluvial soils belong to the Mahidasht, Valiabad, Fakhriabad, and Ganjab series. The Mahidasht Series is the most extensive, developed on alluvium deposited by Ab-i Marik and Qara Su. They are silty clays over clays with *p*H values of 7.7 to 8.5, with carbonate content averaging 32% ($CaCO_3$). Often waterlogged in spring, they quickly dry and shrink to a compact, columnar structure which becomes cloddy under cultivation.

Valiabad, Fakhriabad, and Ganjab series are soils developed on stone-free, silty clay alluvium deposited by transverse tributary streams. Slopes are somewhat steeper than on Mahidasht Series soils but nevertheless do not exceed 1°. None of these alluvial soils presents more than minimal limitations to cultivation in the modern environment. Nevertheless, prior to incision of the trunk streams, which, as argued under Alluvial Chronology and Palaeoenvironmental Influences, probably occurred in the mid-19th century, alluvial plains were likely more prone to spring flooding and were crossed by more tributaries, the dry, meandering channels of which are still evident.

Vegetation

Vegetation History

Sediment cores from the Lalabad and Nilufar spring basins in the study area were raised in 1963 by H. E. Wright, Jr., University of Minnesota. To date, no pollen analysis of these cores has been undertaken, since both show intervals of desiccation of the basins, represented by dark organic-rich (?soil) horizons. Uninterrupted sedimentation was found to have occurred in cores raised in 1963 and 1970 from Lake Zeribar, 160 km northwest of Kermanshah (Pl. 3, inset). A core from Lake Mirabad, 160 km southeast of the town (Pl. 3, inset), showed some late interruption of sedimentation, but cores from both sites were deemed suitable for pollen analysis because of the need to compare vegetation history in these widely separated areas. Preliminary analyses were published by van Zeist (1967) and fuller ones with detailed interpretations were presented by van Zeist and Bottema (1977).

The spectra of pollen types in cores from Lakes Mirabad and Zeribar (van Zeist and Bottema, 1977) are correlated in Plate 3, in which columns have been added to summarize the vegetation and associated climate of pollen assemblage Zones 1–7 (and subzones), and to correlate geomorphic events in the Kermanshah area up to ca. 40 000 B.P. While this area is 160 km distant from both Zeribar and Mirabad, the similarity of vegetation and associated climatic changes in the Holocene epoch common to both cores justifies their use as standards against which environmental changes around Kermanshah can be compared. No purpose is served by further summary of the changes deduced for the Zeribar and Mirabad areas, so the reader is referred to Plate 3 for this information. Pollen analysis of the Lalabad and Nilufar sediment cores would throw important light on the questions of tree cover and of fluctuating ratios of crop-and-weed to grass-and-shrub pollen through the period of sedentary human occupation.

Modern Vegetation

Several millennia of disturbance by grazing, ploughing, and cutting have ensured that very little of what could be called "natural" (unmodified) vegetation remains in the Kermanshah area, or even in the wider region of the central Zagros. The following floristic summary is taken from the work of Zohary (1973).

The area lies within the Kurdo-Zagrosian sector of the Irano-Turanian territory and is dominated by Quercetum brantii in the park forest and by *Artemisia* in the highland steppes (Zohary, 1973:199). The Irano-Turanian steppe-forest comprises two classes: (*a*) Quercetea brantii, in which characteristic species are: *Quercus (persica) brantii, Q. (infectoria) boissieri, Q. libani, Acer cinerascens, Amygdalus communis, A. orientalis, A. horrida, A. kotschyi, A. reuteri, A. salicifolia, A. arabica, Berberis integerrima, Celtis caucasica, Cerasus microcarpa, C. mahaleb, Colutea persica, Cotoneaster racemiflora, Crataegus ambigua, C. aronia, C. monogyna, Daphne angustifolia, Lonicera arborea, Paliurus spina-christi, Pyrus cordata, P. syriaca, Pistacia atlantica, P. khinjuk, and Rhamnus kurdicus* (Zohary 1973:199); and (*b*) Junipero-Pistacietea (Juniperus-Pistacia-Amygdalus steppe-scrub), which "comprises a number of plant associations led chiefly by *Juniperus polycarpos, Pistacia atlantica,* [and] *Amygdalus* spp. . . . It always characterizes marginal conditions for tree- or shrub-growth" (Zohary, 1973:583–85).

Two comments by Zohary (1973) serve to illustrate the effect of human modification of the vegetation:

From Kermanshah . . . the entire route [of a vegetation transect] passes along valleys which are intensively cultivated. Although we are travelling within the climax area of Quercetea brantii, there are scarcely any traces of this Kurdistan forest. . . . Very rarely one meets with stunted remnants of a stand or single trees or shrubs that indicate the former existence of a forest which once covered the slopes. In its stead there is a dense cover of tragacanthic communities on the non-arable lands and an array of combinations of segetal communities on cultivated or semi-cultivated ground [273].

Much of the forest of this class [Q. brantii] has been completely destroyed and converted into cultivated land, in which solitary trees were left for shade. . . . Yet large parts of the . . . [Zagros] ranges still support park-like forests with interspaces between trees ranging from 20 to 50 m. Only in very favourable sites, inaccessible to man, is the arboreal cover fairly high [582].

The geomorphic significance of this area's vegetation lies in the degree of protection from erosion that it affords the surface soil. During periods when potentially erosive runoff occurs (late winter and spring), the cultivated area is either bare or discontinuously covered with grain stubble or fallow-year weed growth. Erosion potential is therefore high. Yet, few signs of damaging sheet or gully erosion are noticeable. Three factors might mitigate high erosion rates. First, but only in the axial zones of the lowlands, surface slope could be too low to impart sufficient erosive power to runoff. Second, the high infiltration rate of precipitation into even fine-textured soils (since these are rarely at field capacity) and shrinkage cracking of fine-grained materials could inhibit surface runoff. Third, when it is generated, surface runoff is rapidly attenuated by diversion into irrigation ditches and ponded by low artificial embankments around and within cultivated fields.

Land Use

According to a recent inventory (AB-AV Khakh, 1973) agricultural land occupies about 1750 km^2 of the study area (allowing for uncertainties as to the coincidence of boundaries used, and the extent of "public lands" in the Kermanshah region). Notwithstanding uncertainties, this figure represents 70% of the combined extent of alluvial lowlands and piedmont fans. West of the Ab-i Marik, however, agriculture is extensive on the pediments not included in these divisions. Of this agricultural area, just over 43% was in wheat and 30% in fallow. Of the wheat acreage, 80% is dry-farmed, with yields of 0.2 tha^{-1}, while the remaining irrigated fraction yields 0.95 tha^{-1} in the Razawar (Sanjabi) Plain, 1.2 tha^{-1} in the Kermanshah Plain, and 3.0 tha^{-1} in the Marik Plain. Numbering over 56 000, sheep are by far the dominant livestock in the system. More than 10 000 goats use the land similarly, usually grazing with the sheep flocks. By contrast there are only just over 6000 cattle (cows, calves, oxen). The important domestic poultry number roughly 10 000.

Modern Settlement Patterns

Modern villages show distinctive patterns of distribution with respect to land types and water sources. In total, 485 villages were counted on the 1:50 000 map sheets (an average of 1 every 10 km^2). In the Ab-i Marik Plain a dozen or so are paired on opposite banks of the river and its tributaries. These were amalgamated for counting. Table 3 shows the percentage frequency of villages over the major landform categories of the area. Note that nearly 72% of villages occur on the alluvial terrace and alluvial fan categories, which cover 53% of the area. The distributions are described in more detail below, using three 1:50 000 map sheets as large representative divisions.

The region around Kermanshah is diverse in land type. It includes the rugged front of Kuh-i Parau, large and small fans at its foot, a narrow alluvial strip flanking the Qara Su, undulating loam-covered hills to the northeast and southwest, and some small, linear rocky ridges. Over the hilly terrain, village density is low, size is small, and distribution is fairly even, although expectably highly concentrated in small cultivable valley floors closer to the water table. In the narrow Qara Su alluvium, villages are distributed evenly along the river, roughly every 2 km–4 km.

TABLE 3. Landform Regions and Distribution of Modern Villages, Kermanshah.

Landform	% of Total Area	% of Villages* (N = 485)
Rock outcrop	7.9	0.3
Colluvial slope	21.0	6.7
Pediment	17.8	11.0
Alluvial fan	29.5	34.7
Alluvial terrace	19.2	37.0
Spring basin/overflow	3.1	6.7
Floodplain	1.4	3.1

*Villages located on boundaries (20.6% of total) were divided equally between adjacent landform categories.

To the southwest, in the middle and upper valley of the Ab-i Marik, the hills southwest of Kermanshah pass through an apron of alluvial fans into a westward-widening expanse of alluvial lowland. Across the river, alluvium gives way southwestwards to undulating fans, pediments, and colluvial surfaces; then abruptly the land rises to the boundary ridge. Villages are distributed in two well-defined belts: one in mid-lowland on both sides of Ab-i Marik, and one in shallow valleys, in the central hills of Kuh-i Sefid and Kuh-i Zengalian and the hill/pediment zone to the southwest. Both village belts clearly take advantage of soil and water resources, although differences in the resources mean that hill-front villages are smaller.

Northwest of this region, the hill/mountain zone encircles the northern end of the study area. On the west side, the piedmont zone is much wider than on the east, but alluvial lowland is the most extensive land type. The lowland is drained by the widening valley of the upper and middle Qara Su, by the Ab-i Khan which enters from the northwest, and by the lower Ab-i Marik as well. In addition, numerous springs rise in the middle of the lowland north and west of the Ab-i Marik, and drain into it. Villages are here distributed in three well-defined belts. One follows the piedmont-lowland transition on the west side, taking advantage of cultivable soils and accessible water at the mouths of shallow valleys dissecting the piedmont. Another follows the Qara Su downstream from Rawansir, based on nearby alluvial soils and water diverted from the large Rawansir spring and the incised river. A third belt is densely concentrated in mid-alluvium along the Ab-i Khan and around the spring basins.

Lastly, in the Razawar valley area, steep-fronted mountains drop to a river- and spring-deposited alluvial lowland east of the river; a bold promontory of mountainous terrain projects northwest-southeast between alluvial zones in the northwest; gently sloping hills border the alluvium in the southwest. Villages are again confined to distinctive belts. One follows the Razawar river; one follows both northeast and southwest borders between uplands and lowlands close to abundant springs and irrigation channels fed by them; and one lies in mid-alluvium both east and west of the river.

Geomorphology

Introduction

A map of physiographic regions in the Mahidasht Project area is included as Plate 4. It was compiled from overlays on 1:21 600-scale airphotos, which were transferred to 1:50 000-scale contour maps, and reduced to 1:125 000-scale. It shows seven terrain types: rock outcrop, colluvial slope, pediment, alluvial-colluvial fan, alluvial terrace, spring basin/overflow, and alluvial floodplain. Individual minor landforms of significance are also shown. The following sections discuss these landforms under three headings: Mountain and Hill Zones, Piedmont Zone, and Alluvial Lowlands.

Mountain and Hill Zones

The northeastern rim of the Qara Su basin consists of a thick thrust sheet of massive, Middle Cretaceous, crystalline limestone, which is internally folded, with dips up to 70°, and sliced by south- and southwest-directed thrust faults (Fig. 11). The discreteness of some structural units within this terrain, as well as erosion of one major and several minor valleys across the structural grain, has divided the highland rim into seven separate, short, northeast-trending ranges with summits ranging in elevation from 2345 m to 3350 m.

The valley of the Ab-i Razawar enters the area through a major break in this rim at Tang-i Kurnawazan, just below 1400 m. Major re-entrants in it occur northeast of Kermanshah at Taq-i Bostan (6 km long, mean elevation 1500 m) and north of Rawansir (also 6 km long, mean elevation 1450 m). The former may be guided by a sag in an anticlinal axis, while the latter is aligned along the southerly continuation of a mapped normal fault.

This massive highland sheds no perennial surface drainage. It is however riddled with subterranean passages and shafts (Waltham and Ede, 1973; Maire, 1978) through which rainfall and snow meltwater is led to the piedmont angle, where it emerges in many perennial springs. The spring line (Pl. 2) marks the outcrop of the basal thrust. Ab-i Razawar, itself possibly guided by a sag in a fold axis, has cut through the limestone thrust sheet into the weaker rocks below and has widened its valley in them.

While all of the drainage from this highland rim is today subterranean, its intensely serrated front points to past episodes of surface erosion. Since fan heads at the piedmont angle are not incised and only rarely manifest evidence of modern deposition, erosion of the dendritically organized ravines appears to have occurred during earlier periods of more intense runoff—either before the subsurface drainageways were developed, or when infiltration was inhibited. The latter case would result from a lowering of the limit of ground freezing in late spring and early summer, as could be expected during a glaciation (Rathjens, 1965; Raynal, 1977). Maire (1978) reported perennial *firn* ("metamorphosed snow") as low as 2000 m in shaded places on Kuh-i Shahu above Rawansir, so it would seem that summer temperatures would have to be considerably depressed for snow to be preserved several hundred metres lower.

The summits of the northeast highland rim, in particular Kuh-i Parau and Kuh-i Shahu, bear remnants of a plateau over which karstic (solutional) topography is intensively developed (Waltham and Ede, 1973; Maire, 1978). Besides providing more gently sloping terrain, these areas preserve a dense grass and shrub cover on *terra rossa* soils, which is heavily grazed by sheep and goat flocks, even at this altitude and at great distance from lowland settlements.

FIG. 11. Rugged front of thrust slice of Cretaceous crystalline limestone rising above Lalabad, 30 km north-northwest of Kermanshah, from 1300 m to 2700 m. Note serrated mountain front with no clear structural control of form, inactive fans at base, and nearly level alluvial surface. 17 May 1978.

TABLE 4. Stages of Karst Development, Kuh-i Parau, Kermanshah.

Stage	Subsurface Activity	Surface Activity
7 (?)late Holocene	Small, unerosive stream. Some stalactite growth and wall fluting. Deposition of stage 2 mud on walls (possibly also in stage 3).	Inactive.
6 (?)early Holocene (warming)	Further reduction in erosion. Partial sediment filling of pot-holes formed in stages 3 and 5.	Shrunken stream keeps cave inlet open.
5 (?)late Würm (cold)	Erosion and trenching of much of stage 4 fill. Last significant erosion phase.	Further incision of surface valley system.
4 (?)mid-Würm (cool)	Deposition of banks of clastic sediment in trench formed in stage 3. In places sediments fine upwards.	Accumulation of clastic fill and stagnation of doline development.
3 (?)early Würm (cold)	Erosion of slots and trenches by much reduced stream.	Incision of surface valley system.
2 (?)Eemian* (warm)	Precipitation of large stalactites and deposition of cave sediments in earlier-formed tunnel.	Accumulation of clastic fill and stagnation of doline development.
1 (cold)	Erosion of large, high-level passage over whole length of system in unsaturated zone.	Main phase of doline solution and surface valley erosion.

SOURCE: Brookes, 1982 (after Waltham and Ede, 1973).
* Eemian Interglacial $\delta^{18}O$ Stage 5e, ca. 125 000 B.P. (Woillard, 1978).

From the observations of Waltham and Ede (1973) on landforms and sediments on and beneath the plateau of Kuh-i Parau, a geomorphic sequence for that karst was tentatively proposed by Brookes (1982), as shown in Table 4.

The southwestern border of the Qara Su basin is formed by a homoclinal ridge at 1800 m–2300 m, with local names for individual segments: Kuh-i Kalehabad, Kuh-i Alivark, Kuh-i Banchal. It is composed of southwest-dipping, Oligocene Asmari Limestone. It manifests a serrated skyline profile due to projection of steep-sided "flatirons" between erosional ravines. The eastern segment of this ridge sheds steep, intermittent streams towards the upper Ab-i Marik. In its west part, west of Shahabad Pass, the drainage divide crosses fold axes in Eocene and Cretaceous formations—Kashkan, Taleh Zang, Amira, and Gurpi—so that the topographic boundary of Qara Su drainage is less well defined. Shahabad Pass itself is developed along an anticlinal axis in the Taleh Zang and Amira formations. Except along the crest of this ridge, bedrock outcrop is masked by a thin talus apron which passes downslope into pediment gravels.

Across the central area of Qara Su basin, between that river and Ab-i Marik, northwest-trending hills are developed on Cretaceous limestones and Radiolarites (formations rich in radiolarian cherty limestones). These rocks are disposed in northeast-dipping thrust sheets and both sets of hills developed on them are bounded to the southwest by steep scarps. The main range is Kuh-i Sefid, at 2000 m–2500 m, on the limestone formations. It continues northwest, across a pass followed by the Shahabad road, declining in elevation to 2000 m–1700 m; plunges into the alluvial lowland of the middle Marik; and reappears to the northwest in the hills which confine that side of the Qara Su basin. The Radiolarite outcrop is broader and the hills developed on it lower (ca. 1800 m) and smoother (Fig. 12), and for the most part mantled with blocky, cherty colluvium.

FIG. 12. Smooth-crested hills at 1900 m on Cretaceous Radiolarites (cherty limestones), mantled with weathered rubble and degraded *terra rossa*, 10 km west of Kermanshah. 10 August 1975.

Piedmont Zone

As the name implies, a piedmont is a transition zone between highland and lowland. The change in slope between them, controlled by either a change in rock type or a fault, influences a change in the regime of highland streams debouching into the lowland. Whether this change of regime in turn causes a change from erosion to deposition will depend on the magnitude of the change in slope, as well as on the power of streams in maintaining erosion across the boundary.

Where the slope break is pronounced, "flashy" streams are intensely erosional in the highland, and produce steep ravines through which the streams carry heavy loads of bouldery and muddy sediment. This sediment is rapidly deposited at the mouths of ravines as fans containing well to crudely stratified alluvium and massive, unstratified, mudflow debris. Where ravines are closely spaced, or where runoff occurs in sheets or rills from an undissected highland front, the debris accumulates in a laterally coalescent apron, known as a *bajada*.

Where the slope break is less pronounced, streams, however flashy, collect more runoff from longer individual watersheds and may therefore be powerful enough to maintain erosion across the slope break. Debouching across it, carrying usually coarse sediment loads, streams here spread laterally in wide, shallow, shifting channels which corrade the piedmont to produce a gently sloping ramp across bedrock, known as a pediment. Declining flows at the end of each flood, or at the end of a period of such floods, leave a veneer of stream-laid gravel over the erosional surface. Where this is thick and bedrock not exposed, the topographic feature may be confused with coalescent, low-angle, alluvial fans.

Both types of piedmont landform are found in the Mahidasht Project area. Contrasts in the geological structure of the southwest and northeast rims of the Qara Su basin are reflected in a broad division into a pediment-dominated southwestern side, and a fan-dominated northeastern side. This division is, however, complicated by the ubiquity of fans downslope from pediments in the former area, and the development of a narrow pediment in the northwest part of the latter area. Further, there is a less easily generalized pattern of fans and pediments flanking the central hills, but this too is recognizably influenced by structural and topographic contrasts within the hills.

Pediments

Pediments are extensively developed along the southwestern piedmont. In the southern part, southeast of Shahabad Pass, Tertiary clastic and carbonate rocks dip southwest and are capped by the thick, resistant Asmari Limestone which builds the boundary ridge. While streams have not deeply dissected this ridge, impermeable clastic formations shed runoff into a dense network of streams which have planed the upturned edges of strata to form pediments 0.5 km–2.5 km wide. The continuity of the pediment ramp is occasionally interrupted by a low, discontinuous ridge developed on a more resistant carbonate formation.

Northwest of Shahabad Pass, subsidiary folds within the Tertiary formations broaden their outcrop and considerably widen the piedmont zone. The boundary ridge on the Asmari Limestone has also been worn back by headward extension of Qara Su basin drainage, so that piedmont watersheds are larger. Streams have consequently been more energetic in planing pediment surfaces 2.5 km–9 km wide, sloping at 1.5° across the Tertiary rocks (Fig. 13). Possibly, two pediments are developed in the north of this zone, because of the interruption of the smooth, wide ramp by a low limestone ridge. This could have acted as a subsidiary divide shedding its own drainage northeast to plane an outer pediment in front of the inner one. Alternatively, inner and outer pediments may be parts of one pediment interrupted by the ridge, as happens to the southeast of Shahabad Pass.

Upslope from Marivani, in the northwest piedmont, a rare exposure in the pediment gravels showed 5 m of crudely bedded, clast-supported, rounded limestone pebble and cobble gravel with imbricated structure (Fig. 14). Further upslope, piles of boulders cleared from ploughed fields showed the clasts in this gravel cover to be larger and not so well rounded, a clear reflection of closer proximity to their source in the Asmari Limestone ridge.

To the south, nearer the distal edge of the pediment

FIG. 13. Two generations of pediments planed across Cretaceous carbonate/clastic sequence, 19 km south of Kermanshah. Upper pediment slopes to right with proximal margin against colluvium-covered hills at far left. Lower pediment slopes to right from bluff cut in distal edge of upper pediment in left centre. Whitish streaks on interfluves express weathered calcrete on pediments. 15 August 1975.

FIG. 14. Coarse, stratified, rounded limestone gravel overlying pediment west of Marivani in the northwest piedmont of the Mahidasht. 23 May 1978.

zone, near Sartappeh, the main road along the west side of the Marik Plain cuts through 2 m of the gravel veneer. The exposure shows an upper zone of strongly indurated, calichified limestone gravel. This has been fractured by weathering into conglomeratic cobbles, often platy in shape, set in a loose matrix of reddish loam. Below the indurated horizon, the gravel has a striking white, powdery matrix of calcium carbonate, with stringers of laminated, cemented carbonate. The base is marked by unweathered carbonate gravels. Similar calcrete profiles were logged in the upper Razawar valley (northeast of the project boundary) and in the valley separating the Kuh-i Sefid and the Radiolarite ridges, southeast of Kermanshah (Fig. 15).

Calcretes have several origins (Netterberg, 1980; Goudie, 1983). These examples have not been examined microscopically, but their parent materials, topographic position, and horizon structure lead to a reasonably confident inference of a pedogenic origin. Soil-water and groundwater had to be available for the solution of carbonates from the parent material, but annual water balance must have been negative for the concentration of reprecipitated carbonate as a cementing agent in the surface horizon. Yet, the regime of precipitation input to the surface could not have been so vigorous as to promote surface runoff—otherwise erosion or sedimentation would have suppressed pedogenesis. Such a regime may be postulated to have marked a change from one of intense precipitation or snowmelt to one with seasons of less intense precipitation, followed by warm seasons. This may be expected to result from a change from maximal to declining phases of glacial cycles in higher latitude zones.

The surface of the calcrete, with its conglomeratic fragments set in a matrix of reddish loam, marks another significant change of surface geomorphic regime. Clearly, the indurated calcrete crust has been disrupted by weathering. The loam, however, similar to *terra ros-*

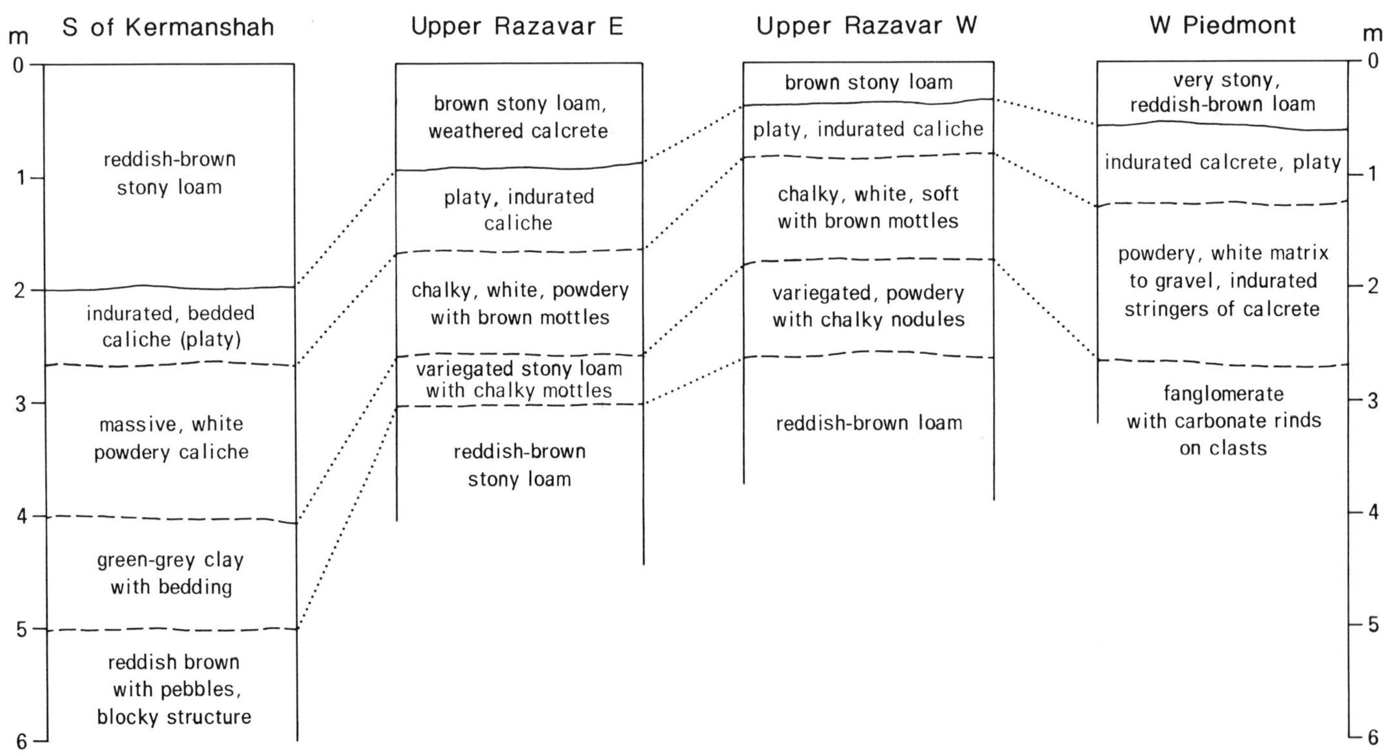

FIG. 15. Profiles of calcretes near Sartappeh (east Marik lowland), south of Kermanshah, and upper Razawar valley (outside project area).

sa widely reported from Mediterranean borderlands, where also carbonate bedrock predominates, cannot so simply be interpreted as a weathering product.

Several studies have shown that the granulometry, clay mineralogy, and quartz ^{18}O ratio of *terra rossa* are more compatible with an allochthonous, aeolian, rather than an autochthonous, weathered origin of the bulk of its parent material (Macleod, 1980; Frolking et al., 1983; Yaalon and Ganor, 1973). This interpretation avoids the difficulty of deriving the bulk of noncarbonate *terra rossa* parent material by solution of carbonate rocks with very low insoluble residues (Macleod, 1980). It also explains the preservation beneath *terra rossa* of micro-weathering relief produced by the corrosive activity of endolithic lichens on carbonate rock surfaces (Danin et al., 1983). This relief could not be preserved on a rock surface being dissolved to produce the *terra rossa*. Further support is provided by measurements of atmospheric dust loadings and aeolian sedimentation in regions peripheral to hyperarid dust sources such as the northern Sahara (Yaalon and Ganor, 1975; Singer, 1967). These are seen as sufficient and necessary to explain the qualities and quantities of *terra rossa* in downwind regions.

Pediment surfaces in the Mahidasht have been dissected by descendants of the streams which formed them. Slope breaks between pediments and valley sides display a prominent white cap of calcreted gravels (Figs. 13, 14), while the valley floors are filled with a reddish, occasionally pebbly, loam. The loam is believed to represent redeposited *terra rossa* from the interfluves.

These loam-floored valleys debouch into the lowland across low-angle alluvial fans in which up to 4 m of stream entrenchment or excavation for road material has exposed up to two units of both gravel and loam. The significance and correlation of these will be dealt with under the heading Alluvial Fans.

Pediments are also well developed in the southeast of the study area, most extensively on the northeast dip slopes of the Kuh-i Sefid and the parallel range of Radiolarites southeast of Kermanshah. Pediments also slope southwest towards the upper Ab-i Marik from the more subdued crest of the southeastern Sefid. Northeast from Kuh-i Sefid, the sequence of terrain types passes through bare rock highland along the crest, into matrix-supported, angular, blocky, limestone rubble with a red loam matrix in the hills (Fig. 16), into clast-supported, rounded, limestone cobbles with an upper horizon of red loam and fewer rounded cobbles over the pediments. Both these latter surface materials are more or less rubbly *terra rossa*, with a preferred pluvio-aeolian origin for the loam matrix. The present valley of the lower Qara Su is incised a little into the pediment surface. The pediment itself, however, slopes uniformly northeast, ignoring the Qara Su valley, while dendritic drainage patterns on its surface are truncated by the valley. Streams which cut the pediment therefore flowed northeast across the present Qara Su so that this valley has developed subsequently, presumably through capture of an earlier northeast-flowing Qara Su by a stream eroding headward from the Gamas-i Ab.

FIG. 16. Colluvium over limestone bedrock, consisting of fresh, angular limestone blocks set in matrix of colluviated *terra rossa*, near Toveh Latif, south-southeast of Kermanshah. 25 July 1975.

On the northeast flank of the upper Ab-i Marik, two generations of calcreted, gravel-veneered pediments slope from the subdued crest of Kuh-i Sefid towards the valley (Fig. 13). Since here there is no reason to believe that bedrock influences the number of pediments, as in the northwestern piedmont (discussed above), a climatically induced change of geomorphic regime is preferred. A climatically based sequence of geomorphic episodes for the pediments is summarized below:

1. Mechanical weathering of bedrock in the steep headwater zones of piedmont watersheds supplied coarse debris to streams fed by intense rains, and possibly snowmelt, during two periods probably correlative with glaciations in areas of higher latitude and altitude. Streams were flashy, braided in pattern, and migrated laterally across the piedmont, planing pediments and leaving veneers of fluvial gravels over them. These two pediment-forming episodes were interrupted by an interval of stream incision which is reflected in the vertical separation of the pediments in the western piedmont and the southeastern part of the central hills.
2. A transition to more equable climate with less intense rainfall and snowmelt and warmer summers led to strong solution of carbonate gravels and reprecipitation of secondary calcite from upward-mi-

grating capillary waters to form a caliche (calcrete) cap in the gravels of the two pediments.

3. Landscape stability was subsequently maintained, but the geochemical regime changed to one in which the indurated calcrete cap was disrupted to produce conglomeratic fragments of calcrete, while airfall dust (from dry and/or wet sources) contributed to production of a reddish *terra rossa* matrix for the fragments.
4. A subsequent period of dissection led to further fragmentation of pediment surfaces into narrow interfluves separated by confluent valleys. These valleys then received the finer matrix of the *terra rossa* and some smaller clasts within it which were eroded from these interfluve areas. The latter were left covered by a veneer of calcrete fragments. Many parts of these interfluves, farther from erosional influences, preserve *terra rossa* intact.

There is no dating control on these events, but a link with the later part of this sequence has tentatively been established with alluvial fan stratigraphy in an exposure west of Mahidasht village, and will be discussed in the next section.

Alluvial Fans

Piedmont fans are ubiquitous and extensive in the Mahidasht Project area, accounting for 29.5% of the area (compared with 23.7% for alluvial lowland). They range in area individually from less than 1 km^2 to nearly 10 km^2, in shape from simple and multiple cone segments to planar aprons of piedmont debris, and in evolutionary history from the product of one runoff event to a complex sequence of gravel and loam units separated by erosion surfaces and soils. Because the balance of these processes has long been in favour of accumulation, few exposures are available from which a regional history of fan development can be deduced. However, an imperfect and tentative history can be discerned from the few sections available, which also permit linkage of fan history with that of both pediments and alluvial lowlands.

Fans occur in two contrasting, but functionally similar, geomorphic situations. Along the northeast highland rim, typical cone-shaped fans debouch from steep ravines incised into the front of the limestone thrust slice which builds ranges such as Kuh-i Parau (Fig. 11). Fans here are steep, due to high topographic relief, and are nested in groups in which small fans from small ravines abut, and are deformed by, larger ones from larger ravines. Where master streams have excavated larger ravines, such as northeast of Kermanshah and north of Rawansir, larger intramontane basins are fringed by small fans which feed larger fans at their mouths.

The northwestern extension of the central hills also contains this type of hill-foot fan, particularly along its northeast flank, and again in intramontane basins, such as northwest of Kermanshah. Along the southwest flank of this part of the central hills (Kuh-i Zengalian), a steep unbroken scarp slope is fronted by a planar apron of colluvial debris (*bajada*) deposited from subparallel chutes which run down the scarp slope. Here, unconcentrated runoff prohibited development of cone-segment fans.

The second geomorphic context of fan development occupies the transition between pediments and alluvial lowlands, predominantly along the western side of the project area. Here, fans have been built by lower-gradient streams emerging from valleys incised into pediments. They are thus low-gradient forms (usually less than 2°) and lateral boundaries between neighbouring fans are indistinct, as are those with alluvium at their distal ends.

Visual examination of the geometry of fans and their contributing basins in this area leads to a generalized and qualitative confirmation of the often-demonstrated quantitative relationships amongst area, length, slope of the fans and their contributing areas (Hooke, 1968; Bull, 1978). These relationships have not been pursued quantitatively here, however, since surface and subsurface evidence indicates that fan development has been complex in two main respects.

First, stratigraphy indicates that fans are composed of superposed units of cobbly gravels and pebbly loams which would probably have had different original surface geometries, which would in turn have had a strong inheritance effect on the evolving form of fans.

Second, alluvial stratigraphy in the lowlands shows that, since the last major episode of fan building, alluvial aggradation has buried the distal ends of fans so that, depending on their slopes, a variable proportion of their original area is missing from the present surface. Fan stratigraphy, then, deserves closer attention than modern morphometry.

Only two exposures were available in alluvial fans, and only one provides an instructive stratigraphic record of its later history. This is located 3 km–4 km west of Mahidasht village, a little inside the border with the alluvial lowland. The section is shown in Figure 17. The basal unit (numbered VI to conform with numbering of lowland alluvial units discussed later) consists of crudely bedded, subrounded and subangular, clast-supported, limestone gravel with pockets of muddy sand matrix. The upper 25 cm–35 cm is impregnated by calcium carbonate, but is not as indurated as the calcrete on pediment gravels. Gravel of this size, shape, and bedding indicates deposition from flashy, torrential, high-bedload streams, which would have risen along the western boundary ridge, probably in the Shahabad Pass. This pass incises the pediment and

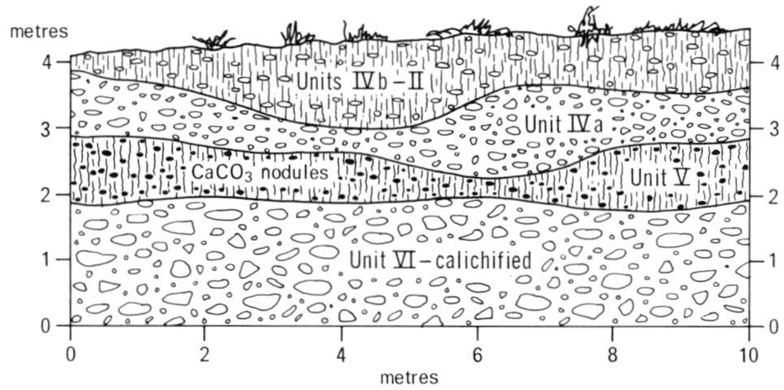

FIG. 17. Section in fan 3 km west of Mahidasht village, showing stratigraphic relationships of sediment Units II–VI.

belongs to that system which dissects the entire western piedmont. The gravel therefore probably is the product of pediment incision. Its age is discussed below under the heading Alluvial Chronology and Palaeoenvironmental Influences.

This calichified gravel is overlain by 0.5 m–2.0 m of reddish mud, with pronounced prismatic structure and powdery calcium carbonate concretions up to 3 cm in diameter. This must have been deposited by moderate runoff and probably had its source in the *terra rossa* on the higher pediments. A similar unit is seen in several cutbanks of streams draining the western piedmont. All exposures of it show similar material and structure to alluvial Unit V ubiquitously exposed in the lower part of trunk-river cutbanks. The prismatic structure and carbonate concretions are pedogenic in origin and speak for a relatively lengthy interval of landsurface stability. The upper surface of this unit is erosional; no evidence of a stable land surface, such as the A-horizon of a soil, remains.

The age of this unit may be bracketed by a tentative correlation of its formative interval with a period of warming climate and increasing tree cover, deduced from the pollen spectra of Lakes Zeribar and Mirabad (Pl. 3) to have lasted from 10 500 to 6000 B.P. (van Zeist and Bottema, 1977). Formation of the well-developed, pedogenic structures may be correlated with the optimum of warmth and maximum tree cover attained at 5500 B.P. (van Zeist and Bottema, 1977). An upper age estimate relies upon correlation of an erosion surface which truncates Unit V, and of a gravel above it, with the same features exposed in many river cutbanks in the lowland, which are argued below to date between A.D. 200 and 1200.

The reddish loam of fan Unit V is overlain, above an undulating erosional contact, by 0.5 m of pebbly and cobbly, subrounded and subangular gravel. Modal clast diameter is 5 cm–10 cm with a maximum of 30 cm. The gravel was clearly deposited by high-energy streamflow or sheetflood across the fan surface from gravel sources in the pediment or upslope on this fan. Its composition, structure, and stratigraphic relations point to correlation with a similar unit exposed in lowland river cutbanks, designated as Unit IVa. It is argued below to have resulted from a brief regional flood, which eroded the loam beneath, incised shallow channels into it, and filled them with this gravel. Chronological evidence suggests a date somewhere between A.D. 200 and 1200.

This gravel unit passes gradationally up into 0.25 m–1.0 m of dark brown, stony loam with the A-horizon of an immature soil on it. This is correlated with up to 10 m of silt/clay alluvium (Units IVb–II) exposed above Unit IVa gravels in lowland river cutbanks. It represents more moderate deposition of finer-grained sediment, with sources in exposures of Unit V revealed by the earlier regional flood, possibly augmented from residual *terra rossa* soils on pediment surfaces. It is apparently not being deposited today, as streams have incised the western piedmont fans, so that its age is bracketed by the age of the flood gravel beneath (between A.D. 200 and 1200) and the mid-19th century, when (as argued below) streams returned to incisive regimes in this region.

Lowland Zone

Alluvial Plain

Alluvium occupies 23.7% of the Mahidasht Project area and is disposed in an N-shaped belt, in which the two arms are formed by the upper and middle Ab-i Marik Plain and the Razawar (Sanjabi) Plain, while the diagonal consists of the middle Qara Su Plain (Pl. 4). Alluvium reaches its greatest extent in the middle and lower Marik Plain where a valley transect is 20 km wide.

The topographic boundary of the alluvium is nearly everywhere against the toes of alluvial fans which border the upland rim and the central hills. This border declines from 1440 m in the upper Marik valley to 1360 m in the middle Marik, to 1340 m in the upper Qara Su valley, to 1300 m in the lower Qara Su. The sharpness of the boundary depends on the slope of fans, so that against gently sloping fan toes (less than 0.5°) the mapped boundary may be in error by as much as 100 m. Geologically, the fan-alluvium boundary is often difficult to distinguish, since the fans, especially along the western piedmont and east of the middle Ab-i Marik, are capped by a thin layer of loam, such as that described above in the fan section west of Mahidasht village. Along the western piedmont, where the youngest fan sediment was brought from piedmont valley floors, the only useful criterion for mapping the boundary of fans and alluvium is sediment colour; fan sediment is usually reddish brown, whereas riverine alluvium is often grey.

Riverine alluvium also borders alluvium deposited

in spring overflow basins and their outlet channels. Extensive spring-laid alluvial basins occur northwest of the lower Ab-i Marik, between the central hills and the middle Qara Su, and in three areas surrounding the Razawar Plain. Around these basins there is no topographic break to assist mapping, and they are often encircled by an irrigation channel which further obscures any natural boundary. The boundary is best identified by (1) the anastomosing or braided pattern of wet or dry channels leading away from springs; (2) wetter, therefore darker, soils in spring-laid alluvium; and (3) alkali efflorescence in these Burbur Series soils.

Sedimentation in the spring basins has probably been from two sources: (1) from overflow channels moving towards the trunk rivers (indicated by lighter toned bars in channels, such as that leading from Nilufar basin to the middle Qara Su), and (2) from standing water ponded behind alluvial surfaces aggraded by the trunk streams (as along the eastern Razawar Plain, where spring basins are noticeably lower than riverine alluvium). This latter cause was proposed by Wright (1968) to account for lacustrine deposits which he cored in the Lalabad and Nilufar spring basins. The existence of either standing water or dry land in these spring basins thus depends to a varying extent on the geomorphic regime of the trunk streams. When these were in an incisive mode, spring overflow channels were free to drain the basins and provide more agricultural land; when trunk streams were aggrading, overflows were blocked and basins flooded. Implications for establishment and permanence of villages in these basins need to be tested against archaeological data.

By far the greater area of alluvium has been deposited by surface-fed elements of the drainage net: first, the trunk streams in the lowland axes—Qara Su, Ab-i Marik, Ab-i Khan, and Ab-i Razawar; second, smaller, now dry channels subparallel to these, especially over the plains of the middle Marik and the Razawar; and third, tributaries rising in the piedmont and joining them at greater angles. Stream-bank sections in alluvium in the first and third of these cases demonstrate that the streams have been responsible for depositing the alluvium bordering them. In the second case, since they are now dry, it can only be assumed that, when flowing, they also deposited some alluvium. The fact that the alluvium thins from up to 12 m thick in the lowland axes to zero against the bordering fans also strongly suggests a riverine deposition. Bank sections also show that the modern lowland surface is underlain by only the youngest of up to four distinctive alluvia. This is normally not more than a metre thick, whereas between 2 m and 12 m of alluvium is exposed.

The alluvial surface shows little evidence of channel migration, such as meander cut-offs, floodplain scrolls, or avulsed channel traces, so it appears that the youngest alluvium has been exclusively deposited from overbank floods, rather than by lateral accretion.

One important exception to this was, however, identified by drawing contours at 1-m intervals from spot heights on the 1:50 000-scale topographic maps. This feature is an abandoned valley branching from the upper Qara Su near Meskinabad (Pl. 4). It is oriented south, then east, and slopes from 1322 m to below 1318 m over a distance of 10 km (average slope 0.004), ending in a blind saddle at 1320 m where it would once have rejoined the Qara Su. In its lower reach the surface of this abandoned valley bears traces of very shallow, anastomosing channels picked out by tonal contrasts on airphotos. These betray torrential runoff in a poorly integrated drainage network, such as might be expected to result from a brief flood event. Arguments made in the following section on alluvial stratigraphy support the occurrence of a regional flood between A.D. 200 and 1200, so this abandoned valley may have been formed by avulsion of the Qara Su during that event. Its currently smooth outlines attest to alluvial aggradation within it since the flood, but this ceased when regional stream incision lowered the main Qara Su channel 3 m–4 m below the avulsion point, cutting off flow through this valley.

While overbank floods appear to have been the dominant mechanism of recent alluviation, there are no levees above modern incised channel walls to betray it. These may have been destroyed by minor bank erosion during incision of the meanders. More likely, they never existed, since the youngest preincision alluvium (Units IVb–II) is predominantly of muddy texture with insufficient volume of coarser particles to settle out close to the flooded channel and form levees.

Another source of alluvial sediment (and indeed of some buried alluvial units) might have been in mud-laden rains. Such events are well known and well studied in the southern Negev and northern Sinai deserts of Israel, and are known to yield much higher proportions of clay than silt-rich dust storms. Sediment sources have been traced to the northeastern Sahara (Yaalon and Ganor, 1973). Extensive areas of southern Israel are underlain by thick sequences of this pluvio-aeolian loess, and its stratigraphic alternation with coarser beds is interpreted as signifying repeated moister (more clay) and drier (more silt) intervals (Issar and Bruins, 1983).

One of these "red rain" events occurred in Kermanshah in mid-May 1978, when a two-hour rainfall from a complete, red-coloured, easterly moving cloud cover yielded about 2 mm of mud. An annual event with this yield of sediment would deposit 20 cm of mud in 100 years. Such events would be expected to be basin-wide, and the mud to be deposited over all land surfaces. Pluvio-aeolian loess as a component of the *terra rossa* of the piedmont and hill zones has been discussed previously, but this deposit is heavily eroded over most areas. While its absence from many such surfaces leads to doubt that air-fall has contributed significant amounts of mud in recent centuries, "cleaner" rains may also have continuously flushed it into valleys.

This supposed loess underlies the small basin of Nahal Besor in northern Negev, Israel, in the general area where it has been studied by Issar and Bruins (1983). Sedimentological studies by Gardner (1977), however, showed this to be a fluvial deposit, so the same genetic argument arises in that area as in the Mahidasht.

The possibility that some of the Mahidasht alluvium had an airborne source is intriguing, yet the only readily accessible sections for study of its structure are in river banks and enough of them show features that indicate fluvial deposition, such as included riverine bivalves and gastropods, pebbly sands, and worn potsherds. Furthermore, where this kind of evidence is lacking, the alluvium shows colour changes between reddish brown and grey, vertically and laterally, suggesting colour differences in watershed materials as the best explanation.

The present courses of the Ab-i Marik and the Qara Su have become incised from 2 m to 12 m into the alluvial plain since the youngest alluvium beneath it was deposited. In the middle Qara Su this incision was briefly interrupted and a narrow alluvial terrace formed in meander loops. The timing of incision will be discussed below, but its cause can be sought in declining sediment yields from the basin, possibly as a result of a more complete cover of erosion-retarding vegetation (itself a likely result of widespread abandonment of arable land). Formation of a narrow floodplain, now stranded as a terrace, could have resulted from a brief return to higher sediment yield, a diminished capacity to transport even reduced loads supplied to the stream, or a briefer, more catastrophic runoff event which could only be accommodated by channel widening.

The channel of the Ab-i Razawar has responded differently to recent regime changes. It has not become incised. Its present bed is markedly contrasted with those of the Qara Su and the Ab-i Marik in being wider (Fig. 18) and heavily charged with mobile gravel and sand from sources in volcanic rocks in its upper basin, outside the project area. Such channels are known to respond to discharge variations by widening faster than they deepen, due to their erodible sandy banks. This behaviour also makes such rivers more unpredictable at flood stage, since waters may avulse into parallel channels, traces of which are plainly evident adjacent to the Razawar.

The incised channels of the Qara Su and the Ab-i Marik for the most part confine all but exceptionally

FIG. 18. Channel of the Ab-i Razawar near Qaqelestan, 33 km north-northwest of Kermanshah. Note wide, shallow, gravel bed with bars; bluff in shadow cut by river at high stages in mounded cultural debris overlying Unit V.

high annual floods. Even that of May 1975, which took out a concrete bridge over the middle Qara Su (but for which no magnitude is available), was so confined, leaving a thin scum of suspended-load mud along channel walls up to 2 m above summer water level. Only in places where stream incision was accompanied by minor lateral migration does inundation occur today.

Modern suspended sediment loads are plotted against Qara Su discharge to provide sediment rating curves at three gauging stations, Doab, Pol-i Khoneh, and Gharbaghestan (Pl. 2 and Fig. 19). The regression coefficients range from 1.29 to 1.80, indicating that suspended sediment load does not increase dramatically with increases in discharge. In the section below on alluvial chronology, unexceptional rates of alluviation over the last thousand or so years are explained with reference to limited access of runoff to sources of fine sediment, as well as to earlier depletion of those sources. These regression coefficients may be explained in a similar way. It may be noteworthy that the lowest rate of increase of suspended load with discharge occurs at Doab. This is the farthest upstream station and is that to which sediment delivery is areally dominated by the Marik basin, in which steep piedmont and hilly zones are farthest from the trunk stream. This would allow for storage of channel sediment in the tributaries and the trunk stream upstream from Doab, which is indeed manifest in channel deposits (Unit I) there. In contrast, the lower two gauging stations, with greater rates of suspended load increase with discharge, receive sediment from nearby steep piedmont and hill and mountain zones with less opportunity for storage of suspendible sediment.

Both the Qara Su and the Ab-i Marik follow highly sinuous courses. Ratios between 2.5 and 4.0 for stream:valley length (sinuosity) are usual. Since sinuosity is strongly and positively correlated to silt-clay content of channel-wall material, it is no surprise to

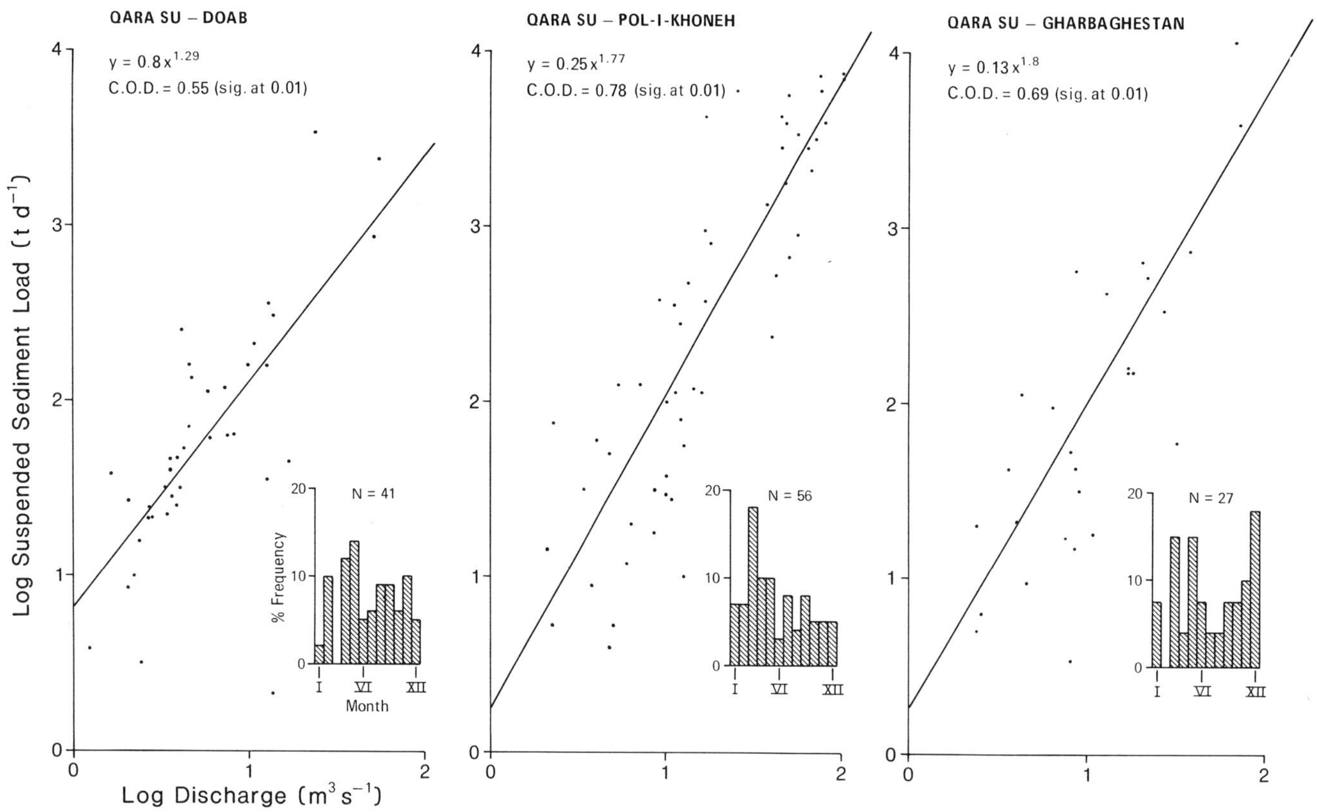

FIG. 19. Sediment rating curves for Qara Su at Doab, Pol-i Khoneh, and Gharbaghestan. Insets show sampling frequency and sample size. C.O.D. = coefficient of determination.

find highly sinuous streams, where bank materials consist of 90% or more silt-clay (see Fig. 27). Meanders are thus prone to frequent cut-off which produces short straight reaches, and this in turn upsets attempts to measure meander geometry in several adjacent loops, since series of regular meanders are rare here.

Again, close inspection of meander geometry reveals a wide range of radius of curvature and of meander shape. Since radius of curvature is positively related to channel width, and width in turn to discharge, the wide range of radii implies an imperfect adjustment of present meander geometry to prevailing discharges. This may reflect a change in discharge levels, or year-to-year fluctuations, which are clear from the annual hydrographs (Appendix) and the annual maximum discharges (Fig. 10).

Explanations of erratic meander geometry are further complicated by the fact that over the past 150 years or so these rivers have been incising their channels into an alluvial plain which they earlier built by overbank flooding. They have therefore inherited a geometry and have been making form adjustments to a changed regime during incision.

In only one location has it been possible to estimate the rate at which meanders have recently migrated. Northwest of Kermanshah, a mid-18th-century fort at Qal-e Khoneh was laid out in rectangular plan. The northeast wall of the fort has been removed by the migrating right bank of the Qara Su which in 1969 (airphoto date) stood 130 m inside the projected line of the fort wall. If it is assumed that the wall originally stood on the river bank, this bank has migrated at an average rate of 60 cm per year since 1750. This is of course a minimum rate, since the fort was probably not built exactly at river's edge.

Alluvial Stratigraphy

The surface of the alluvial lowland described in the preceding section is underlain by mainly fine-grained alluvium, up to 12 m of which is exposed in at least 500 km, and possibly up to 750 km (depending on estimations of sinuosity) of cutbank along the three trunk streams. In addition, many tens of kilometres of tributary stream banks expose more sections. Access to this wealth of information was limited more by the time available and the surface condition of sections than by logistics. Thirty-nine sections were visited and logged (roughly one every 6 km of valley length on average). They range in height from 2 m to 12 m, and are located mainly along the Qara Su and the Ab-i Marik, because the Ab-i Razawar is not so incised.

It was on the nineteenth day of the 28-day reconnaissance study in 1975 that a cutbank was located on the Ab-i Marik, 1 km downstream from Mahidasht village (Figs. 20, 21), which provided the key to interpreting the exposed alluvial record of the project area. The section actually comprises two neighbouring cutbanks of the river. Each of these exposes a different aspect of the stratigraphy, and together they provide the most detailed record of alluvial history. For this reason the section is used as a "standard" with which less complex ones can be compared and correlated. The term "standard" applied to "section" is not used here synonymously with "type section". Five alluvial units labelled I to V and four intervening soils labelled A to D, in order of decreasing age, are schematized in Figure 21, and described as follows:

Unit V. Up to 3 m fine-grained (modal texture, silty clay, ranging from clay to silt loam), reddish-brown alluvium with no sedimentary structures, possibly because of complete impregnation by Soil D. Base not exposed, top erosional in downstream subsection, gradational to Unit IVb upstream.

Soil D. Impregnates entire 3 m of Unit V and presumably extends below section. Structure blocky, with abundant slickensides and Mn stains on block faces. Represents B-horizon of strongly developed soil, with A-horizon and probably top of B-horizon eroded.

FIG. 20. Channel of the Ab-i Marik 1 km downstream from Mahidasht village, looking northeast. Note homoclinal ridge of Kuh-i Zengalian on skyline, meandering channel cut into alluvial lowland; contacts between sediment units and soils show in cutbank to right; modern channel fill of Unit I expressed as side-channel bar. 21 May 1978.

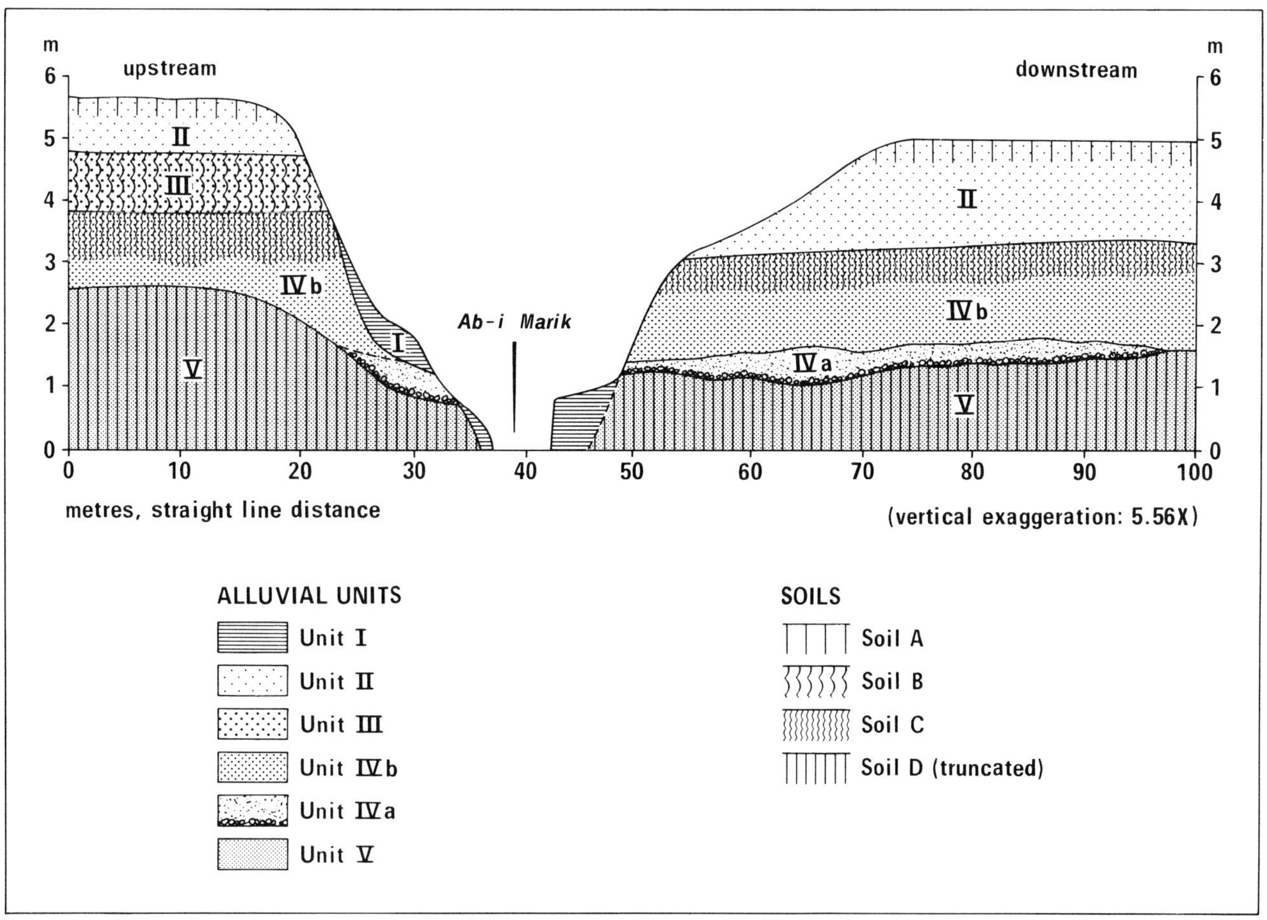

FIG. 21. Stratigraphy of alluvial units and soils at standard section, 1 km northwest of Mahidasht village.

FIG. 22. Downstream cutbank 1 km northwest of Mahidasht village, standard section, lower part, showing blocky-structured Unit V with Soil D, truncated by palaeochannel floor and overlain by Unit IVa gravel. Pick length: 55 cm. 4 August 1975.

Unit IVa. Up to 75 cm of poorly sorted, matrix-supported, subangular and subrounded, limestone and chert gravel with clasts from 1 cm to 5 cm in diameter; matrix of red-brown muddy sand (less than 2-mm fraction contains up to 43% clay). Clasts unoriented. Structureless, but two or three faint partings visible when surface washed (Fig. 22). Contains variably abraded potsherds up to 7 cm long, including abundant plain "Yellow Ware" (Fig. 23) believed to have appeared no earlier than ca. A.D. 200. Also contains bone fragments, bovid teeth, articulated and broken freshwater bivalve shells (Fig. 24). Deposit floors a wide, shallow palaeochannel cut across Unit V (Fig. 25). Unit IVa absent from above Unit V, upstream.

Unit IVb. One and a half metres of upward-fining silty sand to clay silt alluvium; structureless, rare potsherds sporadically distributed. A coarser base may represent decline of flow which emplaced Unit IVa in palaeochannel, but most of it represents later, repetitive alluviation over an abandoned channel and the flanking floodplain. Upstream section shows Unit V passing up into Unit IVb through a 15-cm transition of lightening colour with structure changing from blocky to massive.

Soil C. Impregnates upper 75 cm of Unit IVb. Moderately developed Ah-horizon over weakly developed Bca-horizon. Present only in upstream section.

Unit III. Absent from downstream section. Upstream, 1 m of massive silty clay alluvium.

Soil B. Strongly developed through all of Unit III with thick, strong Ah- and Bca-horizons.

Unit II. One metre of silty clay alluvium, structureless, friable. Forms top of section.

Soil A. Twenty centimetres Ah- (?Ap-)horizon over unaltered C-horizon, presently ploughed and grazed.

Unit I. Discontinuous, variably textured modern channel bars in mid- and side-channel positions.

This sequence records changes in the balance between erosional, depositional, and stable regimes of the channel and floodplain surface at this location. Unit V is believed to represent overbank aggradation of alluvium derived from fine-grained, reddish loams (*terra rossa*) formed and reworked earlier within the piedmont zone. A link between lowland and piedmont occurrences of this unit is provided in a fan section 3.5 km west of the standard section described above (Fig. 17). There, a unit with texture, structure, and colour comparable to Unit V also underlies unsorted gravels below an erosional contact, although in the fan the unit is impregnated by $CaCO_3$ nodules. These may reflect better subsurface drainage beneath a sloping fan than in a floodplain. Upslope from this fan, as in all fans along this western piedmont, valleys incised into pediments are floored with reddish loam derived from pediment interfluves between them, as previously discussed.

FIG. 23. "Yellow Ware" potsherd in Unit IVa, standard section, 1 km northwest of Mahidasht village. Matchbox length: 5 cm.

FIG. 24. Bovid tooth and articulated riverine bivalve shell in Unit IVa, standard section, 1 km northwest of Mahidasht village. 4 August 1975.

FIG. 25. Floor and wall of palaeochannel cut in Unit V, floored with Unit IVa gravels (figure rests pick on these), and filled with Units IVb–II muds. Right bank of unnamed tributary of the Ab-i Marik, 3 km northwest of Mahidasht village.

Soil D on Unit V is thick enough (greater than 3 m) to indicate a relatively long interval in which alluviation had either ceased or was slow enough to permit intense pedogenesis. That this did not produce a calcic B-horizon (unless it was eroded) may suggest either a more humid climate or more acidic soil solutions than in the recent past, when younger soils developed such horizons in a relatively short interval. Another inhibiting factor may have been the relatively carbonate-free sediment load of the river which, as inferred above, was derived from carbonate-free *terra rossa*.

The disconformity between Units V and IVa, its form as a palaeochannel, the absence of an A-horizon on Soil D, and the texture, structure, and sherd content of Unit IVa, all converge to indicate that this long interval of pedogenesis was abruptly closed by an erosive, short-lived flood. The formerly quiescent Ab-i Marik incised its channel, then rapidly filled it with unsorted, structureless, muddy gravels, fining upwards into silty sands. There is no evidence here, or in any other section in which it was encountered, to suggest that this unit was deposited in more than one flood, which probably lasted less than a week.

An equally abrupt return to low-energy, overbank alluviation of muddy sediment is recorded in Unit IVb. The fact that, at the upstream subsection, outside the palaeochannel, no disconformity is seen between Units V and IVb (Fig. 21, left) requires explanation. The magnitude and extensiveness of the flood which produced Unit IVa would appear to have required an erosional contact between Units V and IVb on the surrounding floodplain. Certainly, the A-horizon of Soil D is missing there and the differences in structure between the units indicate a change in conditions. The only viable explanation would seem to involve an early resumption of ploughing on the eroded Unit V surface and its continuation through the first years of Unit IVb's accumulation. Ploughing would have mixed the two units and masked a contact.

Units III and II and Soils C, B, and A (Fig. 26) mark subsequent changes in the balance between alluviation and pedogenesis. When the former was faster, soils did not form; when slower, pedogenesis impregnated the alluvial surface.

Deposition of Unit II (but not pedogenesis of Soil A on it, which continues today) was halted by down-cutting of the Marik. Subsequently, aggradation has resumed, as witnessed by the stabilizing bars represented by Unit I.

No section in the area, other than the "standard" one described above, shows as complex a sequence of alluvial units, disconformities, and soils, but a majority of them, particularly above the lower Qara Su, show the same three basic elements: (1) a basal unit (Unit V) of reddish-brown, muddy alluvium impregnated by a soil (Soil D) with blocky structure, sometimes with $CaCO_3$ nodules, which is truncated by an often-channelled erosion surface; (2) a gravelly or sandy unit (Unit IVa) above this, usually containing potsherds; and (3) one or more younger muddy alluvial units (Units IVb–II) intercalated with up to three soils (Soils C, B, and A), more or less strongly expressed.

The textural character of several samples of each of these units, with three from alluvial fans, and four modern suspended load deposits, is represented in Figures 27 and 28. Figure 27 is a simple sand-silt-clay triangle which clearly distinguishes the sandy flood deposit of Unit IVa from all finer-grained deposits. The

FIG. 26. Upstream cutbank, standard section, 1 km northwest of Mahidasht village, showing stratigraphy of alluvial units and soils. 5 August 1975.

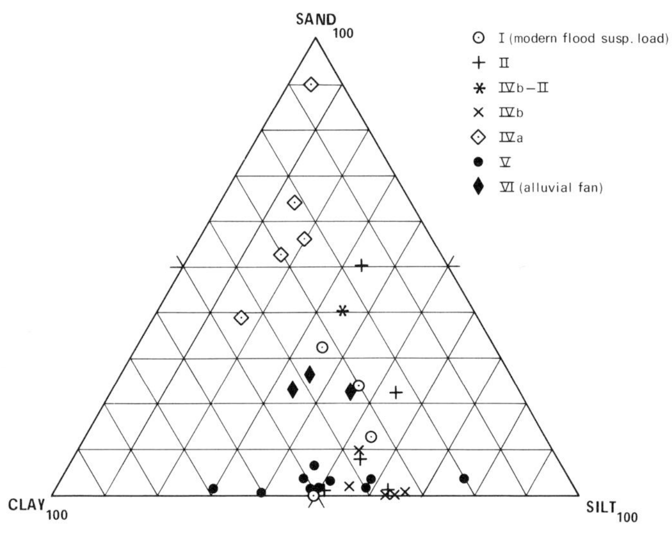

FIG. 27. Triaxial plot of sand, silt, and clay percentages in samples of alluvial Units I–V.

alluvial fan samples plot in the general field of riverine alluvium since it is only the approximately 2-mm fraction of the matrix of fan deposits which was analysed. Their lower sand and higher silt content reflects shorter-lived flows and deposition of less well sorted sediments. These fan samples are closely grouped, with 22%–27% sand and 32%–42% clay. The closest grouping is that of Unit V (sand content consistently less than 10%, clay 20%–70%) with the field of Unit IVb (not more than 10% sand, 32%–42% clay), the field of some Unit II samples, and one modern suspended load (1975 channel-wall deposit). The textural similarity of Unit V and Unit IVb samples may reflect their close chronological association, originating respectively before and after Unit IVa flood gravels. Unit V deposits would have formed the main source of Unit IVb sediment during landscape recovery. The consistently low sand content of Unit V is in harmony with its hypothesized origin as redeposited *terra rossa*, largely of aeolian origin from distant sources.

A more dynamic display of textural parameters can be seen in Figure 28. This plots the diameter of the coarsest 1% (C), a measure of the competence of the transport medium, against median diameter (M; 50% coarser and 50% finer), a measure of the overall texture of the sediment. Again, Unit IVa flood gravels are clearly distinguished, with C values of 300 μm–50 000 μm and M values of 10 μm–10 000 μm. The variability of CM plots of Unit IVa attests to a variety of piedmont source areas, distances of transport to lowland rivers, and hydraulic conditions during its deposition. Significantly, these points plot closest to Passega's "pattern V" (river bed load).

Unit IVb, as muddy overbank alluvium, shows a much tighter grouping, with C values from 200 μm (fine sand) to 63 μm (silt) and M values in the silt/clay range. Unit II, of similar origin, shows more variability, with C and M values ranging into sand, a possible reflection of sources more variable than those of Unit IVb (its variable colour also betrays this); more vigorous depositional processes are not in evidence.

CM plots of 10 Unit V samples show M values in the silt/clay range, as would be expected from Figure 27. The wide range of C values (eight in the range of medium sand and finer, two in the granule class) is instructive in pointing to fluvial rather than aeolian transport to the loci of deposition. That all but one sample have M values finer than Passega's "pattern III" (quiet water) can be interpreted as evidence of floodplain slack water deposition of sediments with a predominantly muddy source; that is, piedmont *terra rossa*.

Lastly, modern alluvium CM plots overlap all others but Unit IVa, supporting the conclusion that they share an origin in overbank deposition by swollen trunk streams.

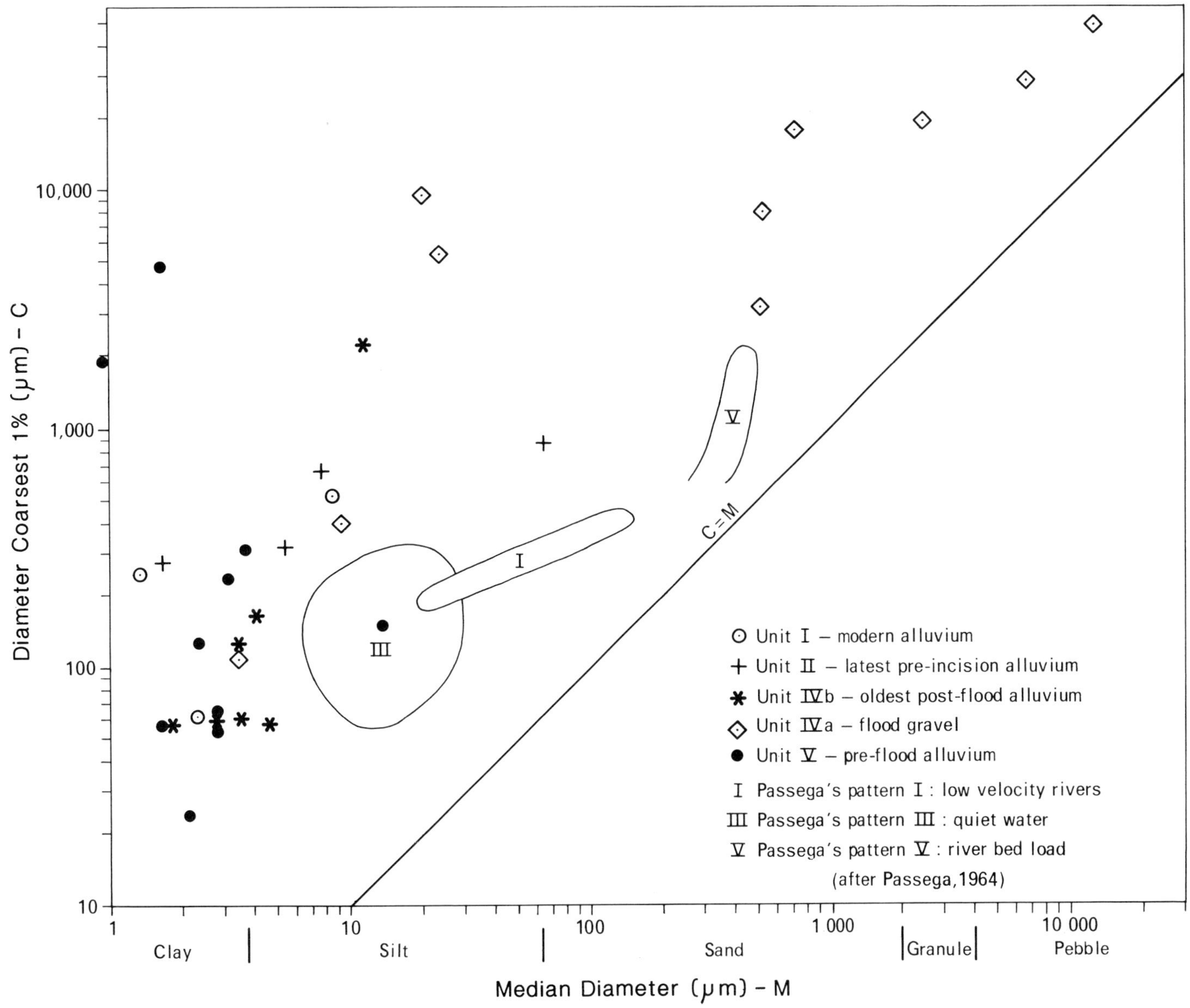

FIG. 28. CM plot of samples of alluvial Units I–V.

Alluvial Chronology and Palaeoenvironmental Inferences

The episodes or events responsible for the geomorphic sequence inferred in the preceding section were regional phenomena and require regional explanation. This is attempted as part of the following discussion of chronological evidence, which, it should be stressed at the start, is not at all secure. The firmest evidence is archaeological, and, while it relates only to the later part of the sequence, it is that part which is more important for interpretation of archaeological data. Earlier parts of the sequence are assigned to shorter or longer intervals, according to degree of development of soils and nature of the sediment. They are tentatively correlated with vegetation changes recorded in the pollen spectra from Lakes Zeribar and Mirabad (Pl. 3). The chronology is anchored more or less securely at four points: (1) the end-Pleistocene ca. 14 000 B.P., (2) the mid-Holocene, ca. 5500 B.P., (3) A.D. 200–1200, and (4) mid-17th century A.D.

Unit VI. This unit is exposed in only two sections, both in fans—one west of Mahidasht and one northeast of Kermanshah. It is a cobbly fan gravel heavily impregnated with powdery $CaCO_3$ nodules and carbonate rinds on clasts. It is interpreted as the product of torrential floods rising in the higher piedmont and surrounding mountains, where both older pediment gravels and mechanically weathered bedrock provided abundant sources of debris. Its place in the stratigraphic sequence (Fig. 17) and the degree of calichification lead to its assignment to end-Pleistocene time. Pollen spectra in the Lake Zeribar core, Zone 3, are interpreted as reflections of a cold, dry steppe, radiocarbon-dated at 33 500–14 000 B.P. (Pl. 3). Abundant evidence of Late Würm descent of the limit of glaciation in the Zagros Mountains is interpreted to reflect temperatures 4°C–7°C cooler than today (Brookes, 1982). Such cold and dry climates would be conducive to mechanical bedrock weathering and flashy spring snowmelt runoff. Calichification of these gravels would have been possible upon cessation of high-energy deposition, as snowfall decreased and temperatures increased through the transition to the Holocene (Pl. 3).

Unit V. This unit overlies Unit VI in both fan exposures available, and it forms the lower part of river-bank sections along the Ab-i Marik, along some of its left-bank tributaries from the piedmont, along the Qara Su to near the Ab-i Razawar confluence, and in one section on the Razawar. Continuity of Unit V between lowland and piedmont is established, and a source in *terra rossa* loams has previously been argued, as has deposition from moderate runoff—probably sheetflow on the fans—and overbank floods from the trunk rivers and tributaries over the lowlands.

Moderate flows reflect moderate rains and snowmelt, as well as regulation of runoff by vegetation. Pollen evidence from Lakes Zeribar and Mirabad shows a gradual increase of tree pollen in Zone 5, dated at 10 500–6200 B.P., which is inferred to reflect warming climate (Pl. 3). Unit V is thus assigned approximately the same age.

In a section on the middle Qara Su near Berim Vand, Unit V occupies the lowest 1.5 m above low-water level. It has a muddy sand texture and its structure comprises individual beds dipping at a low angle (ca. 10°–15°), which identify the deposit as a point-bar sequence. Fragments of freshwater bivalve shells from the sediments yielded a radiocarbon date of 7620 B.P. ±140 (QU-395). Such riverine shell material is known to give anomalously old dates, due to incorporation of older carbon from previously decayed organic matter (Keith and Anderson, 1963), but how much "too old" in this case is impossible to determine.

Soil D. The thick, truncated Soil D on Unit V is inferred to reflect a relatively long interval of landscape stability. While alluviation of Unit V need not necessarily have entirely ceased, the thick soil indicates that it had slowed considerably to permit pedogenesis to penetrate more than 3 m. These conditions imply minimal sediment yield from the basin, which in turn implies either very moderate runoff or maximal vegetation cover, or both. Pollen Zone 7 in the Lake Zeribar core contains rapidly increasing oak pollen which reached a maximum ca. 5500 B.P. (Pl. 3).

A sediment core from Nilufar spring basin (15 km west-northwest of Kermanshah) showed 1.25 m of organic lake mud (gyttja) with a radiocarbon date of ca. 5000 B.P. at the base. This overlies a black organic clay believed to be a marsh palaeosol, and this in turn is underlain by several metres of clay and marsh sediment, with a radiocarbon date of 28 000 B.P. at the base (H. E. Wright, Jr., pers. comm., 1977). Thus, lacustrine sedimentation appears to have been continuous from at least 28 000 B.P. to just after 5000 B.P., when the basin dried up temporarily. Since containment of water in spring basins relied upon alluvial aggradation around them, this appears to have ceased or slowed in mid-Holocene times in the middle Qara Su, north of the Nilufar basin. This would have permitted development of Soil D, which is thus inferred to have begun ca. 6500 B.P. It was abruptly truncated by the event which formed Unit IVa, discussed next.

Unit IVa. This unit is a poorly sorted, usually structureless, matrix-supported gravel, containing potsherds, as well as other "cultural" debris (bone, teeth), and riverine bivalve and gastropod shells. It is inferred to have been deposited during one brief flood which affected the entire watershed. Evidence for its age comes from several sources which are discussed below in order of increasing reliability.

1. A radiocarbon date of 3520 B.P. ±95 (I-9821) was obtained on articulated, riverine, bivalve shells from the unit at the standard section. More reliable evidence from the age of sherds in the unit indicates that this date is, however, too old by at least 2000 years. This is explicable with reference to contamination with "old" carbon-14.
2. At Khorammabad-e Olia, 6 km south of Rawansir, an obviously recent small alluvial fan was aggraded across the border between an older fan and an alluvial terrace. Where its surface is ploughed it is littered with small, abraded sherds most likely turned up by ploughing. None was archaeologically diagnostic, but one small sliver of blue glass, indicative of Islamic technology (post–A.D. 640) tentatively provides a maximum age for the fan. Its significance lies in correlation of this fan with Unit IVa gravels exposed in Qara Su cutbanks nearby.
3. At Qaqelestan, 33 km north-northwest of Kermanshah, the Ab-i Razawar has eroded a 12-m cliff in the north side of a large village mound. The lowest cultural debris of the mound rests on a sterile, blocky structured, silty clay alluvium above a sharp contact, about 2 m above low-water mark (Fig. 29). This alluvium is clearly correlative with Unit V exposed in many sections along the Ab-i Marik and the Qara Su. It is dark grey rather than reddish brown in colour, since it would have been derived directly or indirectly from mafic volcanic rocks in the upper Razawar basin.

FIG. 29. Section in left bank of the Ab-i Razawar at Qaqelestan, 33 km north-northwest of Kermanshah (see also Fig. 18). Prominent break at X marks contact of lower Unit V (with Soil D on it) and overlying cultural debris of village mound, late Parthian in age. 13 July 1978.

Here the upper 30 cm of this Unit V is noticeably darker than material below and contains mud-filled tubules, which are either root or worm-burrow casts. Organic carbon was determined at 0.28%, but is only 0.10% in the lower material. These features tentatively identify this darker horizon as the A-horizon of Soil D on Unit V, which was here preserved by the overburden of cultural debris. The lowest horizon of this debris contains sherds which are no older than late Parthian (A.D. 1–226), so the flood which eroded Soil D at all other sections occurred later than this period.

4. Potsherds recovered from nearly all exposures of Unit IVa in the area, though usually worn and nondiagnostic, include many with a lightweight, yellow fabric, known informally as "Yellow Ware" (Fig. 23). This ware is common on abandoned village mounds; it first appears in the Sassanian period (A.D. 226–640), and continues to the present. Unit IVa must postdate manufacture of this ware, and thus cannot predate A.D. 226.
5. Finally, one "Yellow Ware" sherd from Unit IVa at the standard section has been dated at A.D. 975 ±120 by thermoluminescence (Archaeometry Laboratory, Oxford University, 1976). Acknowledging uncertainties in this dating technique, and the lack of corroborative dates on more sherds from the unit, at least this date does not conflict with other chronological evidence. It suggests a date after ca. A.D. 850 for Unit IVa.

Discounting the anomalous shell radiocarbon date, the remaining four lines of evidence converge rather than conflict, and point to an age for Unit IVa of no older than A.D. 220 and possibly as young as post–A.D. 1000. How much later than A.D. 1000 Unit IVa was deposited is constrained partly by the time required subsequently to deposit up to 10 m of alluvium and form up to three soils, and partly by historical evidence for cessation of this later deposition, discussed next.

Units IVb–II and Soils C, B, and A. At Mahidasht, 1 km from the standard section, the foundations and approach road of a mid-17th-century A.D. bridge have been buried by alluvium of undivided Units IVb–II, and later partially re-exposed by incision of the Ab-i Marik (Fig. 30). Alluviation of these units, beginning immediately after deposition of Unit IVa, thus continued past the date of bridge construction and ceased when the river began to incise. The cause of incisive behaviour in rivers is normally to be sought in conditions that reduce sediment yield from a watershed. Under such conditions, flood discharge is able not only to remove from a channel reach all the sediment brought into it from upstream but to scour the channel bed as well. In the post-17th-century period, only the middle decades of the 19th century can be seen as a time when such conditions might have existed. It was then that a period of government extortion of landowners, possibly combined with the effects of a plague year in 1830 and devastating floods in 1832 (Lambton, 1980), led to noteworthy economic decline in Kermanshahan province. This may have led to depopulation, abandonment of arable land, an increased cover of weeds and grasses, and a consequent reduction in soil erosion. Streams would thus change from aggradational to degradational regimes.

The end of deposition of the alluvial Units IVb, III, and II is thus tentatively placed at ca. A.D. 1850. If the beginning is placed at the earliest date which does not conflict with any of the other evidence for the age of preceding Unit IVa (i.e., A.D. 850), these later alluvial

FIG. 30. Bridge over the Ab-i Marik at Mahidasht village, showing mid-17th century A.D. foundations buried by alluvial Units IVb–II, partially re-exposed by recent stream incision. 11 July 1978.

units have accumulated, at least along lowland axes, at an average rate of 0.2 cm per year in the upper Ab-i Marik, and at 1.0 cm per year in the middle and lower Ab-i Marik and parts of the middle Qara Su. These are minimum rates, for two reasons. First, Unit IVa could be as young as mid-13th century A.D. if the blue glass from the correlative fan is from the Seljuk period, and if speculation is correct that the Mongol invasion was a cause of landscape change (see following section). Second, the allotted interval does not allow for formation of Soils C, B, and A as alluviation periodically decelerated.

Even acknowledging the subdued development of Soil C, and the immaturity of Soil A, traditional concepts of soil development rates would hardly, in the time available, seem to allow for the maturity of Soil B, with its strongly calcic 50-cm B-horizon. A model of pedogenic carbonate development on aggrading floodplains, which alleviates the time problem, has, however, been put forward by Leeder (1975). According to Leeder's diagrammatic summary (fig. 2, p. 264), with an alluvial accretion rate of between 0.2 cm and 1.0 cm per year, as in the Mahidasht, a 50-cm calcic horizon can develop in much less than a thousand years.

However accurate or inaccurate they may be, these alluvial sedimentation rates are calculated for specific small areas around well-studied exposures and should not be applied widely over the alluvial lowland. Almost nothing is known of the stratigraphy or thickness of alluvium away from the trunk rivers, except that it thins to zero against the alluvial fans. The rates are high compared to values determinable from less closely constrained data given by Adams (1965) for the Diyala basin to the west of Kermanshah, and by Kirkby (1977) for the Deh Luran Plain of Khuzestan. Adams quotes 8 m of alluvium accumulated since ca. 3200 B.C. (or 1.54 mm per year), while Kirkby quotes an alluviation rate of 1 mm per year from 8500 B.P. to 4000 B.P. These are, however, in the lower reaches of much larger drainage basins in which much sediment can be stored upstream. Also, nothing is known from those areas about intervals of nondeposition which, if longer than depositional ones, would raise these rates.

Data on thickness of deposits from individual overbank floods on American rivers range from 0.008 ft to 0.114 ft (0.24 cm to 3.42 cm), while data for the recurrence interval of overbank flooding provide a mean value of 1.54 years (Wolman and Leopold, 1957). If a period of a thousand years is allotted to aggradation of Units IVb–II in the Kermanshah area, 650 such events would be expected. If each deposited the range of thickness given above, they would have yielded a total of 1.6 m–22.2 m of alluvium, which encompasses the thickness of those units.

The relief and climate of the Mahidasht would lead to expectation of higher aggradation rates. Lower yields may be due to (1) prior removal of fine sediment from the watershed to yield alluvium of Unit V; (2) further removal as an effect of the devastating flood which deposited Unit IVa; and (3) armouring of river beds, particularly in upper reaches, with gravels brought down in that flood, and consequent protection of deeper alluvium from channel erosion.

Channel incision which followed deposition of Units IVb–II has now ceased, as witnessed by the accretion of mid- and side-channel bars in the main streams, which are composed of several upward-fining units of sand, silt, and clay. This accretion is most pronounced in the upper and middle Ab-i Marik (Fig. 20) and Qara Su where they are fed by a denser network of tributaries crossing a narrower alluvial lowland. There is, thus, less opportunity for storage of recent sediment along these tributaries than along those which cross wider, level tracts.

This recent accretion probably marks watershed response to increasing population since pacification of the wider region under Reza Shah Pahlevi (1877–1944) in the years immediately prior to World War II. In the last 25 years or so, mechanization of agriculture and intensified grazing of sheep and goats owned by formerly transhumant, but now more sedentary, peoples have further increased sediment yield and accelerated this aggradation.

The sequence of geomorphic events on the floodplains of the Mahidasht Project area is summarized in Figure 31. The lower part of the diagram shows changes of floodplain elevation resulting from changes in the balance between sediment yield and runoff over the watershed. Where the curve rises, more sediment is being delivered to the channel than runoff can remove. The channel bed is thus aggraded and overbank flooding with resultant floodplain sedimentation ensues. Where it flattens, alluviation has slowed so that pedogenesis is more effective over the floodplain and soils are developed. Where it falls, a drastic change in runoff regime has lowered the floodplain surface. Only in the last 200 years has channel incision (dashed line) been significantly greater than floodplain lowering. The upper part of Figure 31 shows the chronology of Units I–VI, Soils A–D, the caliche (Ca) on Unit VI, and episodes of river incision, all of which are correlated with floodplain changes by vertical lines.

The foregoing section of this report adds considerable detail to the records described and interpreted by Vita-Finzi (1969) at 12 geological sections in north and northwest Iran, one of them apparently on the Ab-i Razawar near its confluence with the Qara Su. In that study, Vita-Finzi recognized two alluvial units. "Teheran Alluvium" (after Rieben, 1955), consisting of poorly sorted gravel and sand, and assigned a Late

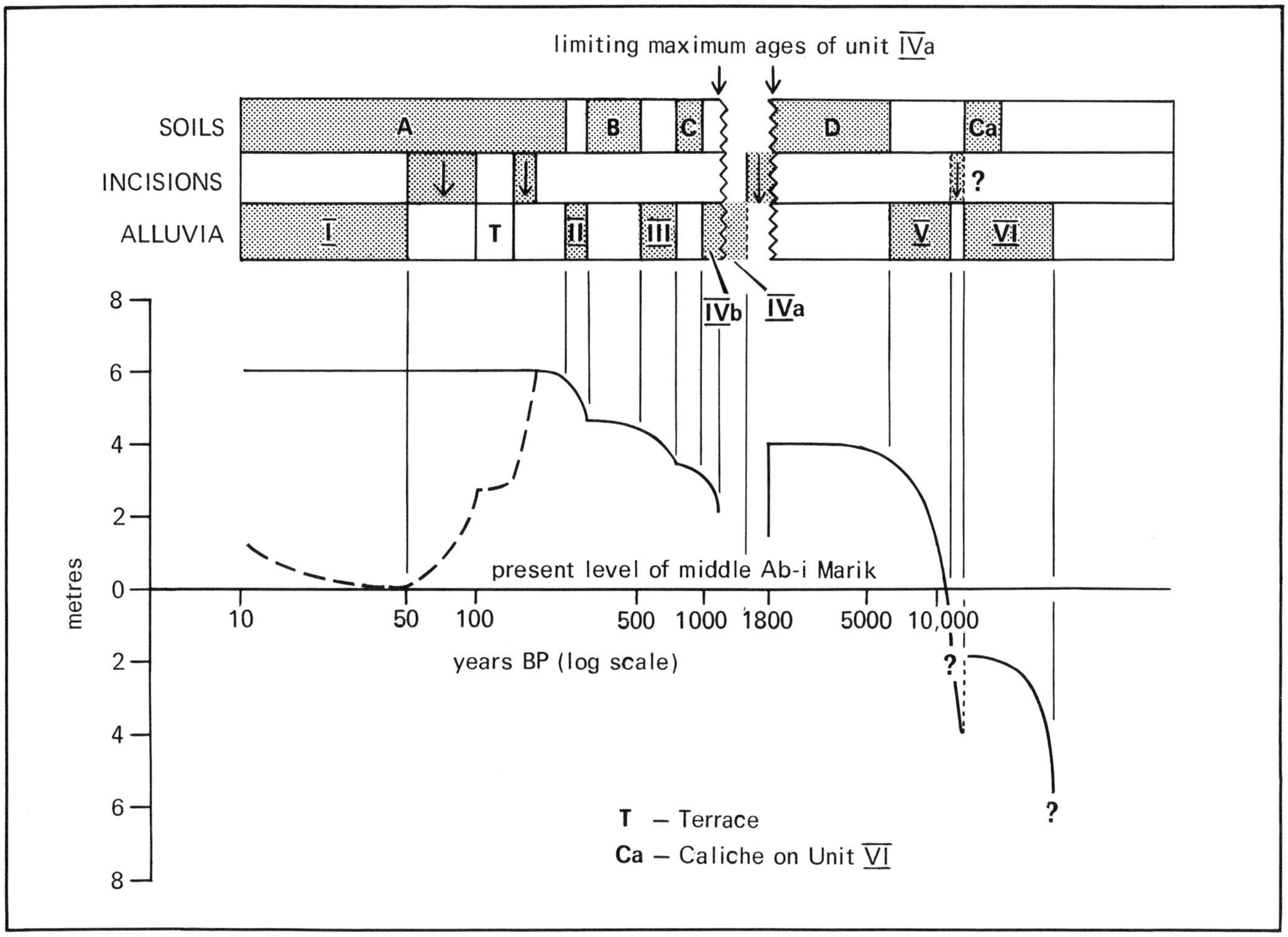

FIG. 31. *Lower:* Plot of floodplain height in the middle Ab-i Marik against time (log scale); dashed line represents channel incision since Unit II deposition. *Upper:* Chronology of alluvial units, soils, and periods of stream incision, correlated with floodplain development stages by vertical lines.

Pleistocene to Early Holocene age, was attributed to a cold-climate, irregular, flashy stream regimen. "Khorramabad Alluvium," a better-sorted, finer-grained unit, assigned to medieval times (ca. A.D. 1200–900), was attributed to more regulated flows carrying finer loads from more vegetated basins, because of increased cyclonic rainfall (possibly associated with circulation change accompanying the "Little Ice Age").

The latter half of the Holocene is thus largely unrepresented in this scheme, although the age of the disconformity between the alluvia was not inferred.

Vita-Finzi's sequence shows some similarity, but more significant differences, to that reconstructed herein for the Qara Su basin around Kermanshah. First, the change from Unit VI (calichified Late Pleistocene fan gravels) to Unit V (redeposited *terra rossa* muds with a thick Soil B-horizon) occurred in the end-Pleistocene as temperatures, effective moisture, and ground cover increased. Change of stream regimen thus occurred earlier than in Vita-Finzi's scheme.

Second, the latter half of the Holocene, which is missing in Vita-Finzi's scheme, is herein largely represented by the thick soil (Soil D) on Unit V.

Third, a major disconformity also occurs in the present scheme, but its origin is firmly tied to a brief flood about a thousand years ago which was responsible for gravelly Unit IVa. In Vita-Finzi's scheme, the disconformity (at least in his Kermanshah-region section) may be correlative.

Last, rather than lithostratigraphic units being assigned wholly climatic causes, in the present scheme the entire Holocene is seen as dominated by muddy fluvial sedimentation, albeit with variable rates, punctuated by longer and shorter pedogenic intervals. The only high-intensity fluvial activity was confined to a few days about a thousand years ago—the brief flood, which, in this context, must be seen as a "catastrophic event."

The Mahidasht Flood—Speculation on Causes

There is firm evidence that Unit IVa was the product of a brief regional flood which truncated the deep soil (D) developed on Unit V, and which transported gravels several kilometres across the lowlands to trunk streams. These first incised their beds by a metre or more, then laid down the gravel in wide, shallow channels. No event so devastating to the landscape had occurred since the late glacial phase of cold, dry climate in which the Unit VI fan gravels were deposited, nor is there any evidence that anything approaching it in magnitude has occurred since. Chronological evidence places this event in the period between A.D. 850 and at least several centuries before A.D. 1850. One slim piece of evidence would place it after A.D. 1220.

The simplest explanation of this flood would call upon a large precipitation event with very large recurrence interval (low frequency). Such events, with return periods possibly as long as five thousand years, have been documented in the United States on geological evidence by Costa (1978a, 1978b). Costa (1978a) documented a minimum recurrence interval of a thousand years for the Colorado Big Thompson River flood of 1976, which was caused by intense rains from a deep low-pressure area stalled for 24 hours over mountainous terrain. Certainly, winter meteorological patterns over the Zagros Mountains would not prohibit such a development in this area, which is in fact closer to Mediterranean moisture sources than eastern Colorado is to Pacific sources.

Statistical considerations, however, cast doubt on a long-return-period event. The probability (r) of a flood's exceeding one of return period (T) in a specified future period of time (P) is calculated using the following formula:

$$r = 1 - \left(1 - \frac{1}{T}\right)^{P}$$

If T for this case is given a value of 1000 years and P is 1000 years, r = 0.63. That is, if the Mahidasht Flood is assigned a return period of 1000 years (it occurred about a thousand years ago) and P is the present (1000 years after the flood), there is a 63% chance that such a flood has occurred in the last 1000 years. If T is given a value of 5000 years (such a flood had not occurred in the previous 5000 years) and P is 1000 years, the probability that a flood has occurred in the last 1000 years rises to 87%. In other words, the larger the return period of an event, the more likely it is to occur when it has not occurred for a long time.

Doubts thus raised lead to consideration of a shorter-return-period event, the effect of which was enhanced by watershed conditions which had radically changed. It is tempting in this area to ponder human activity as potentially lowering the threshold of watershed response to an intense precipitation event. Assignment of the flood to the period after A.D. 850 and long before A.D. 1850 prompts consideration of a cultural discontinuity which could have transformed the landscape, thereby lowering its response threshold.

Western Iran lay at the hub of conflicts amongst Arabs, Turks, Iranians, and Mongols between the mid-11th century, when the Turkish Seljuks took Baghdad, and the mid-13th century, when the Mongols, under Chingiz Khan, sacked it. In 1197–98 Kermanshah province "was laid waste by the . . . amir Miyadjuk" (Lambton, 1980:180) from Khwarizm (astride the lower Oxus) during his rout of the Seljuks from the region. Then, in 1220–21, Miyadjuk's dominions were overrun by the Mongols, under Chingiz Khan (Lambton, 1980). While the rapacity of the Mongol conquest of lands between the Oxus and the Mediterranean may have been exaggerated, the effects of conquest of those lands have surely not. Lambton (1980) cites an account of the Arab historian Ham'd Allah Mustawfi, which refers to Kermanshah's having been reduced to a mere village as well as to declining production in Kurdistan as a result of the Mongol conquest. Several reports of population reduction by Mongol slaughter and capture, as well as details of catastrophic economic decline all over Iran, are referred to by Petroshevsky (1968). Even without deliberately destroying a delicately adjusted system of irrigation agriculture, the Mongols would certainly have had little interest in its maintenance. Historical evidence, however, indicates that several centuries before the Mongols reached what is now Iraq, Seljuk administration had deteriorated to a point at which finely adjusted irrigation systems had probably fallen into disrepair, so that "the Mongol invasion seems to have dealt no more than a coup de grâce to an urban civilization whose roots in rural agriculture had already withered" (Adams, 1965:74, in reference to the Diyala basin). However caused, the demise of irrigation systems in the Kermanshah area would have rendered both natural and artificial drainage channels incapable of accommodating runoff from a high-magnitude storm.

Archaeological Implications

The history of the end-Pleistocene and Holocene landscape of the Mahidasht Project area outlined in the previous section must now be considered in relation to the interpretation of archaeological survey and excavation data. These data are currently incompletely reduced and conclusions about the cultural sequence cannot yet be firmly drawn. The following discussion must therefore focus on the implications of landscape history for the observed morphology and distribution of archaeological sites. As noted in the introduction, the geomorphological study was originally conceived as potentially providing information of value to the interpretation of the cultural sequence. It has shown however that environmental changes through the Holocene were gradual and subtle, and that the most dramatic shift since latest Pleistocene time occurred in the last thousand years. While it is not known at present what effect the shift had on human populations, this event and subsequent environmental recovery have greatly affected the available surface record of the majority of occupations.

Preflood Environments

Following deposition of cobbly fan gravels around the lowland edge in latest Pleistocene time, climates warmed and dried to induce calichification of the gravels. Subsequently, increasing warmth and surface moisture through the early-to-mid-Holocene (Pl. 3) was reflected by tree cover increasing to attain "parkland" density by 5500 B.P.

Alluvial Unit V has been assigned to this early-to-mid-Holocene interval on stratigraphic grounds. Its generally muddy texture and reddish-brown hue attest to fine-grained sediment yield from source areas in the piedmont, under hydrological conditions of moderate rains with equally moderate runoff to aggrading lowland rivers with low banks. Soil D signifies a long period of pedogenesis dominant over alluviation from about 6500 years ago to about 1000 years ago, when it was brought to an abrupt end by the flood that deposited Unit IVa.

The environment of the study area through all but the last millennium of the Holocene epoch can thus be seen as a benign, passive influence on human activities and their distribution. In this environment human settlements would unarguably have been as finely adjusted in their distribution with respect to soil and water resources as they clearly are today. As shown in Table 4, modern villages are highly concentrated in the alluvial lowland (43.7%) and on alluvial fans (34.7%, predominantly on the low-angle, finer-textured fans outside the northeastern piedmont); 18% occur in the combined "upland" landform categories of pediment, colluvial hills, and rock outcrop. Of 515 archaeological sites recorded in the 1975 survey, 44.5% are in the alluvial lowland, and 37.5% on the alluvial fans. Only 10.4% of sites occupy the higher, more rugged landforms. This deficiency of upland sites compared to modern villages may be seen, at least partly, as a reflection of upland site destruction by the Mahidasht Flood.

The record of village mounds several thousand years old in the lowland, however, attests to their survival there against erosional modification. This is significant in relation to a conclusion reached by Kirkby and Kirkby (1976). They observed slope profile changes on mounds of differing ages in Khuzestan and interpreted them to mean that mounds would not survive more than two thousand years of erosion in that semiarid climate. Survival of many mounds in the Mahidasht abandoned before 3200 B.C. might reflect a moister climate and denser ground cover here, but the original conclusion may also be in error, since age is only one of several variables that influence the slope profiles of village mounds.

Effects of the Mahidasht Flood and Subsequent Alluviation

When we come to the effects of the event that deposited Unit IVa, and to the subsequent history of at least the lowland landscape, the conditions of site preservation become less favourable. Mention has been made of the comparative underrepresentation of mounded archaeological sites in the uplands and of recent erosion as a possible contributor to this. In detail, where preserved, these sites occupy topographic positions where erosion would be hindered, such as level and convexly sloping areas where runoff would be slow or diffused; they also occupy small areas outside small piedmont watersheds from which, similarly, runoff would be small. In less protected areas on steeper or concave slopes, and closer to drainageways, severe flooding would ensure erosion of many sites. Since the gravels of Unit IVa can have originated only in the piedmont zone, they bear witness to the severity of piedmont erosion.

Equally if not more significant for archaeological site distribution are the effects of alluvial aggradation of Units IVb–II over the lowland and the lower parts of fans and piedmont valleys. The absence of the A-horizon of Soil D and the depth of the palaeochannels incised into Unit V attest to a small amount of lowland surface erosion during the flood. Some erosion of mounded archaeological sites may also have occurred. Surprisingly few, however, show evidence of gullying, although this may have been obliterated by subsequent human activities. Any record of peripheral slopewash deposition lies buried beneath later alluvium.

The accumulation of up to 10 m of this alluvium has, however, had far greater effects. Here, we are faced with site burial rather than erosion, for which there are two kinds of evidence. The first is stratigraphic, and comes from the exposure in river banks and in one excavated site of cultural material in mounds below the level of the surrounding plain. For example, site no. 468, Jameh Shuran, 4 km east-southeast of Mahidasht village, is a mound standing 5 m above plain level, with cultural material in the top metre dated between 200 and 400 B.C. The age of material 1 m below plain level is estimated at 1000 B.C. The site was thus established before the alluvial plain was aggraded to its present level, and was abandoned long before the medieval flood. It was originally high enough to remain projecting through the alluvium later accumulated around its lower slopes. A pit dug to 140 cm near the break in slope from mound base to plain level revealed only structureless brown silty clay, with occasional sherds and stones, but no evidence of material eroded by slopewash from the mound during the flood. If this material is present, it must lie lower, and would indicate that postflood alluviation has attained a minimum thickness of 1.4 m.

Further evidence of site burial can be found in the distribution of sites mapped by size and age, which suggests that small sites, and even larger ones abandoned before the flood, now lie beneath the alluvial surface. This hypothesis was tested by mapping all sites containing a component of cultural material of known early age in relation to the extent of alluvium in the lowland. Sites containing pre–Bronze Age material (earlier than ca. 3200 B.C.) were chosen for this analysis because this material originally accumulated on sites established on the uneroded surface of Unit V on which Soil D had only just begun to develop (Pl. 3). The sites are sufficiently old for detection of size and side-slope decreases due to both "normal" and "flood" erosion of sites with no younger material on them. Further, they are old enough for the presence of later material on many to be used as a measure of site growth against recent alluvial burial. They are also widely enough distributed over lowland and piedmont zones to permit differentiation of geological effects on site size within and between these zones.

Sites with pre–Bronze Age material in the surveyed part of the Mahidasht Project area, as well as the extent of alluvium, are shown in Plate 5. Sites abandoned before the Bronze Age (i.e., Neolithic and Chalcolithic sites containing no later material) are almost all restricted to the area outside the alluvium or lie close to its border with fans, where it would be expected to be thinnest. Two of these sites occur well within the alluvial border, near the modern road across the Ab-i Marik, but these are only 0.5 m and 0.2 m high and so could represent only the unburied summits of higher structures.

Sites with either major or minor pre–Bronze Age components show no such geographic restriction. These were occupied, if only intermittently, in the Bronze Age or later, and occur in a range of heights up to 8 m within and outside the alluvium. Because they were occupied longer, they were likely higher before the flood than older mounds, and therefore less susceptible to burial by postflood alluvium. Also, middle and late Islamic occupations would have raised originally smaller mounds above the rapidly rising postflood alluvial surface. That pre–Bronze Age material is found at all on sites in this second category attests to the recycling of some of that material by later occupants. Similar distribution maps of sites abandoned at selected periods since the Chalcolithic would more rigorously test the burial hypothesis, but at this time analysis of sherds collected from over a thousand sites has not progressed sufficiently to provide this information.

Notwithstanding this deficiency, the implications of recent alluvial aggradation for interpretation of site distribution are important. For example, the marked concentration of early sites along the alluvial border could, without the geological information, be seen as an adaptation of village location to complementary agricultural resources; with it, this can be seen as a geological artifact.

Postflood alluvial burial has affected not only the distribution of lowland sites predating the flood, but the size of individual sites as well. Observable heights of these mounds represent only their unburied portion (Fig. 32), so that, depending on their original height, average slope, and the thickness of alluvium, the proportion of the original mound now exposed will vary greatly from mound to mound. Geometric relations amongst these variables are derived in Figure 33a and are plotted in Figure 33b–d for original mound heights of up to 10 m; slopes of 2°, 5°, and 10°; derived original areas up to 100 ha; and alluvial thicknesses of 2 m, 5 m, and 10 m.

FIG. 32. Abandoned village mound (*tepe*) at Cheqa Narges, site no. 039, 10 km west of Mahidasht village. Modern village clustered around base. Mound height: 22 m, but several metres of cultural debris lie below plain level. 5 August 1975.

Note that the ratio of original to present mound area is proportional to the square of the ratio of original to present height $(h_1 \div h_2)^2$. The effects of alluvial burial on decreasing mound area can be illustrated with two examples. A mound sloping at 2°, with an original height of 5 m and partially buried by 2 m of alluvium, has a present area of 6.45 ha, but had an original area of 9.29 ha. A mound with the same slope, an original height of 10 m, buried by 5 m of alluvium, now has an area of 6.45 ha; its original area was 25.8 ha, or four times as great.

Of course, few mounds grew by simple conical accretion with circular outlines and constant side slopes,

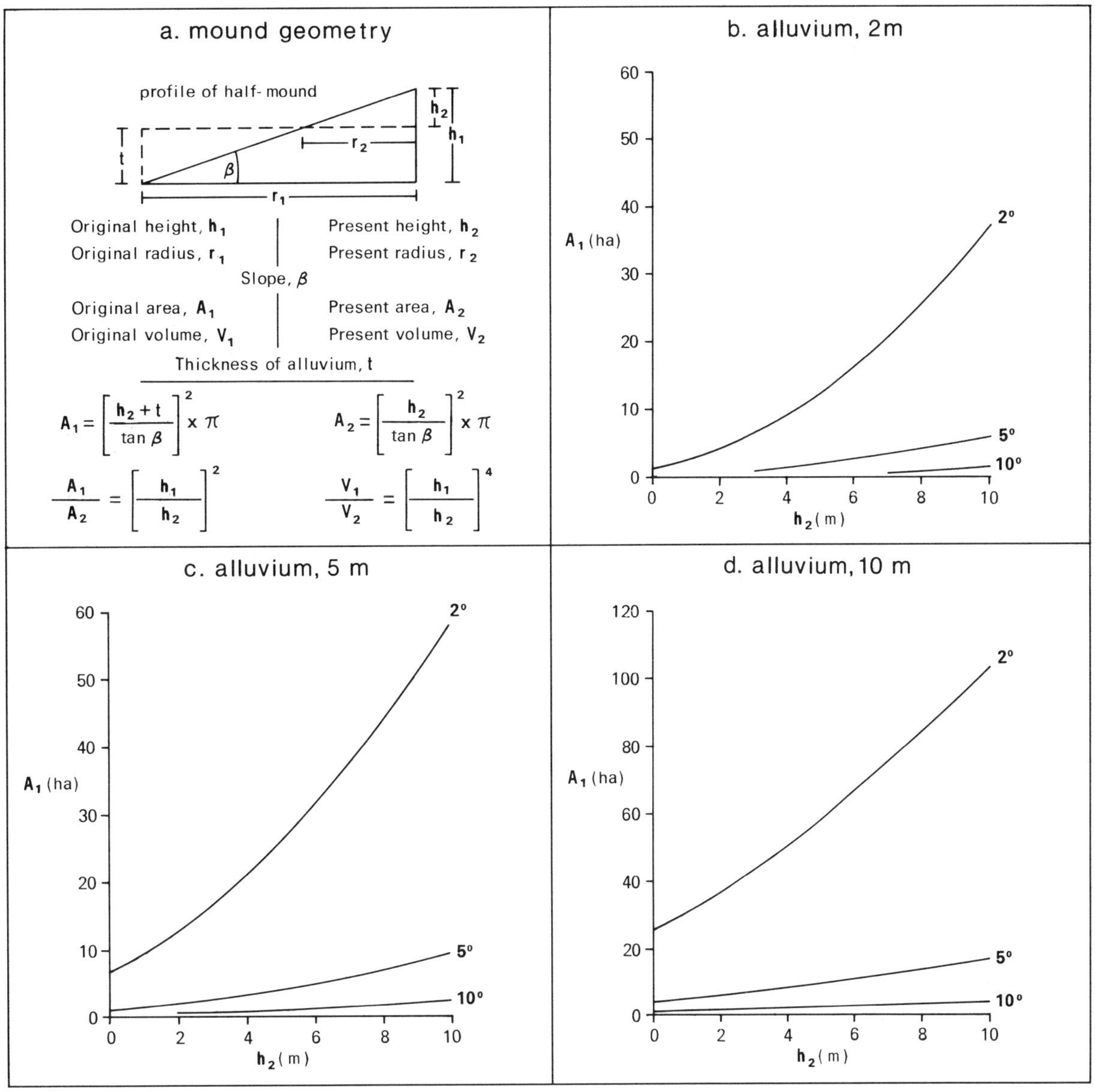

FIG. 33. Calculation of mound size: *a*, simplified geometry of mound and alluvium, with derivation of original area (A_1) and volume (V_1) from present height (h_2), slope angle (β), and thickness of alluvium (t); *b–d*, original mound area (A_1) as a function of present height (h_2) and slope angle (β = 2°, 5°, and 10°), with alluvial thickness of 2 m, 5 m, 10 m, respectively.
Note: Vertical scale of *d* is different from *b* and *c*.

but the implications remain at least qualitatively the same. In fact, the more the geometry of mound accretion departed from a simple growing cone, and consequently the gentler the side slopes and lower the height, the greater the reduction in area by alluvial burial of its base.

The present height of a mound above a topographic datum could theoretically be affected by subsidence of the cultural structure into the alluvial foundation, due to the increasing load of the mound as it grew. This will not affect conclusions drawn on the distribution of pre–Bronze Age sites, since any subsidence would have been complete long before the deposition of the post-flood alluvial Units IVb–II. The same applies to sites established and abandoned in much of the time before the flood. Cultural additions since the flood, which have raised mound summits above accumulating alluvium, could conceivably have led to further subsidence, but this cannot have amounted to more than a few centimetres (less than or equal to 1% of mound height). Since the error in determining mound height is larger than this, any such effect has to be ignored.

Although site area is commonly used both as a measure of site population and to rank sites within a hierarchy of settlement functions, it should be clear from the above that recent alluviation has rendered such analyses meaningless in the Mahidasht. What now appear to have been hamlets could in fact be partly buried villages, apparent villages could be towns, and towns small urban centres. The effects of burial are just as significant for estimates of the bulk of cultural material within mounded sites, since the ratio of original to present mound *volume* ($v_1 \div v_2$) varies with the *fourth* power of the ratio of original to present heights $(h_1 \div h_2)^4$.

These conclusions unfortunately cannot be converted into quantitative archaeological data for two reasons. First, time, labour, and technology can hardly provide sufficient knowledge of the growth history of a representative sample of mounds for their geometric relations with alluvial history to be tested. Second, we do not have sufficient information on the thickness of alluvium over the lowland, although a uniform decrease from axes to margins is suggested by the limited evidence.

These limitations aside, Finnie (1979) was able to support the broader implications of the burial hypothesis in a detailed study of a 500-km^2 transect across the middle Marik, from which 237 archaeological occurrences were recorded in the 1975 and 1978 surveys. For example, he showed that no Chalcolithic sites (ca. 5000–3500 B.C.) occur within 2 km of the present Ab-i Marik, where alluvium is between 3 m and 6 m thick. Also, he found that surface sherd scatters older than middle Islamic were markedly underrepresented within the alluvium and attributed this to burial by Units IVb–II. Older scatters are restricted to more stable landsurfaces above the 1360-m contour.

Finally, the recent geological history of the Mahidasht Project area has rendered comparisons between modern and ancient environments of occupation equally meaningless. The regional flood severely eroded the piedmont, stripping finer soils down to rock or gravel, and therefore depleting the agricultural resources.

In the lowland, prior to the flood, geomorphic activity was minimal, agricultural resources optimal. In the interval since, the lowland first rapidly aggraded, potentially causing problems such as silting of irrigation channels, ponding of water in backswamps, and consequent salinization. Since the mid-19th century, stream incision has decreased flooding, made irrigation water from channels less accessible without pumping, and has lowered water tables, desiccating many smaller stream channels and reducing flow to wells.

These two effects of recent geological history—modification of site size and distribution, and spatial environmental change—render all attempts at placing the archaeological data in their environmental context little more than crude at best.

Appendix

Hydrographs of daily discharge (m^3s^{-1}) of Qara Su at Gharbaghestan, and bar graphs of daily precipitation (mm) at Kermanshah, 1958–59 to 1969–70 water years.

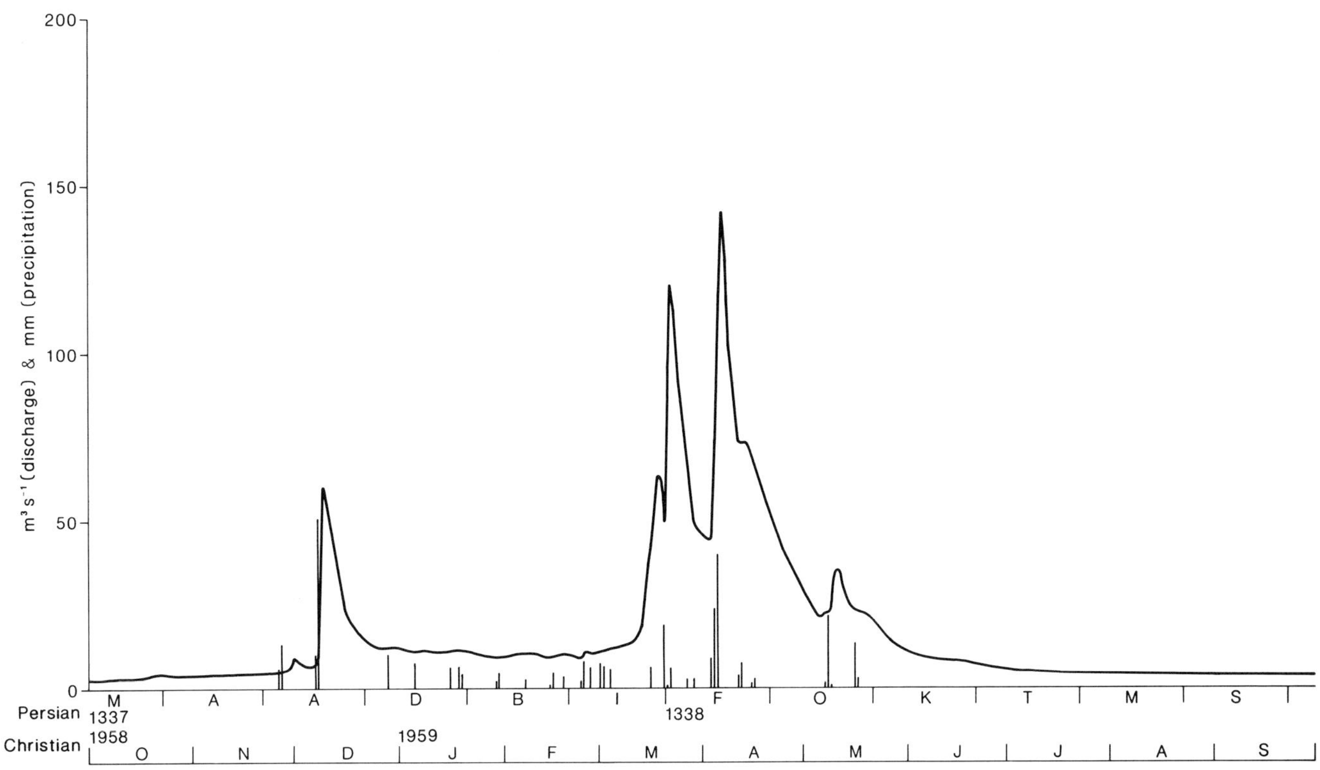

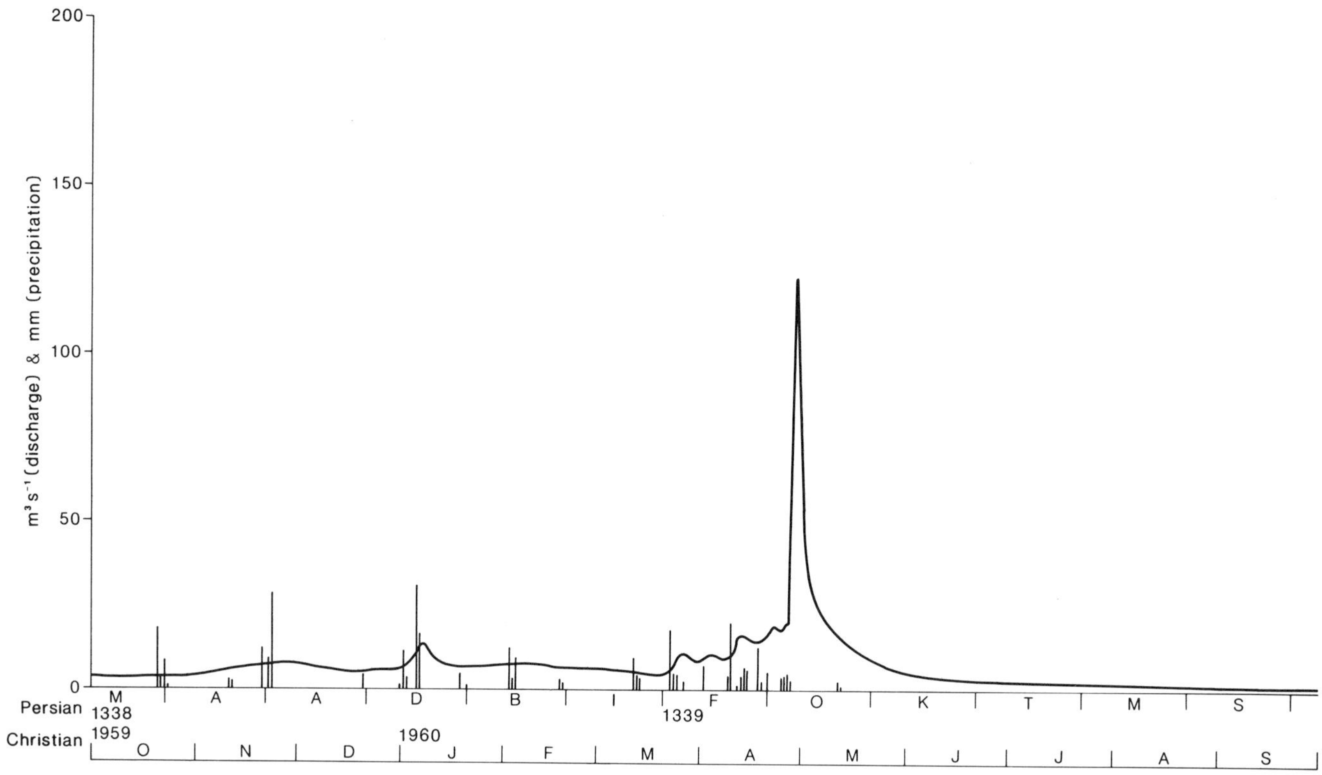

200
150
100
50
0
m³s⁻¹ (discharge) & mm (precipitation)
Persian
M A A D B I F O K T M S
1338
1339
Christian
1959
1960
O N D J F M A M J J A S

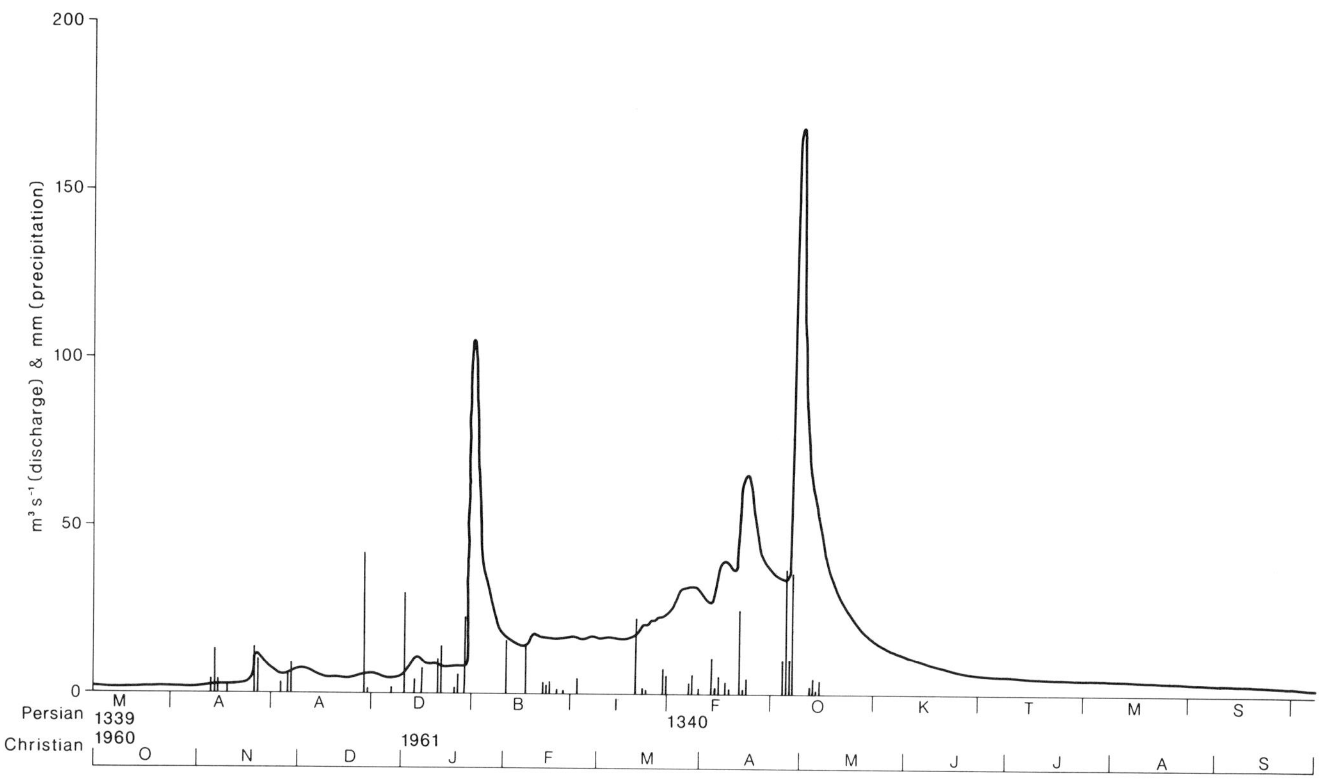

200
150
100
50
0
m³s⁻¹ (discharge) & mm (precipitation)
Persian
M A A D B I F O K T M S
1339
1340
Christian
1960
1961
O N D J F M A M J J A S

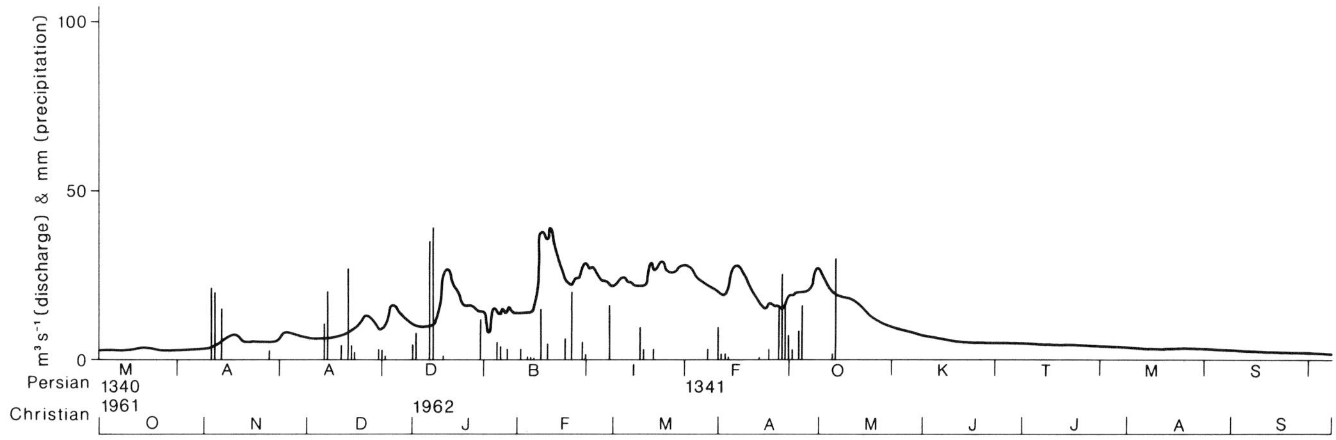
m³s⁻¹ (discharge) & mm (precipitation)
100
50
0
M
A
A
D
B
I
F
O
K
T
M
S
Persian
1340
1341
Christian
1961
1962
O
N
D
J
F
M
A
M
J
J
A
S

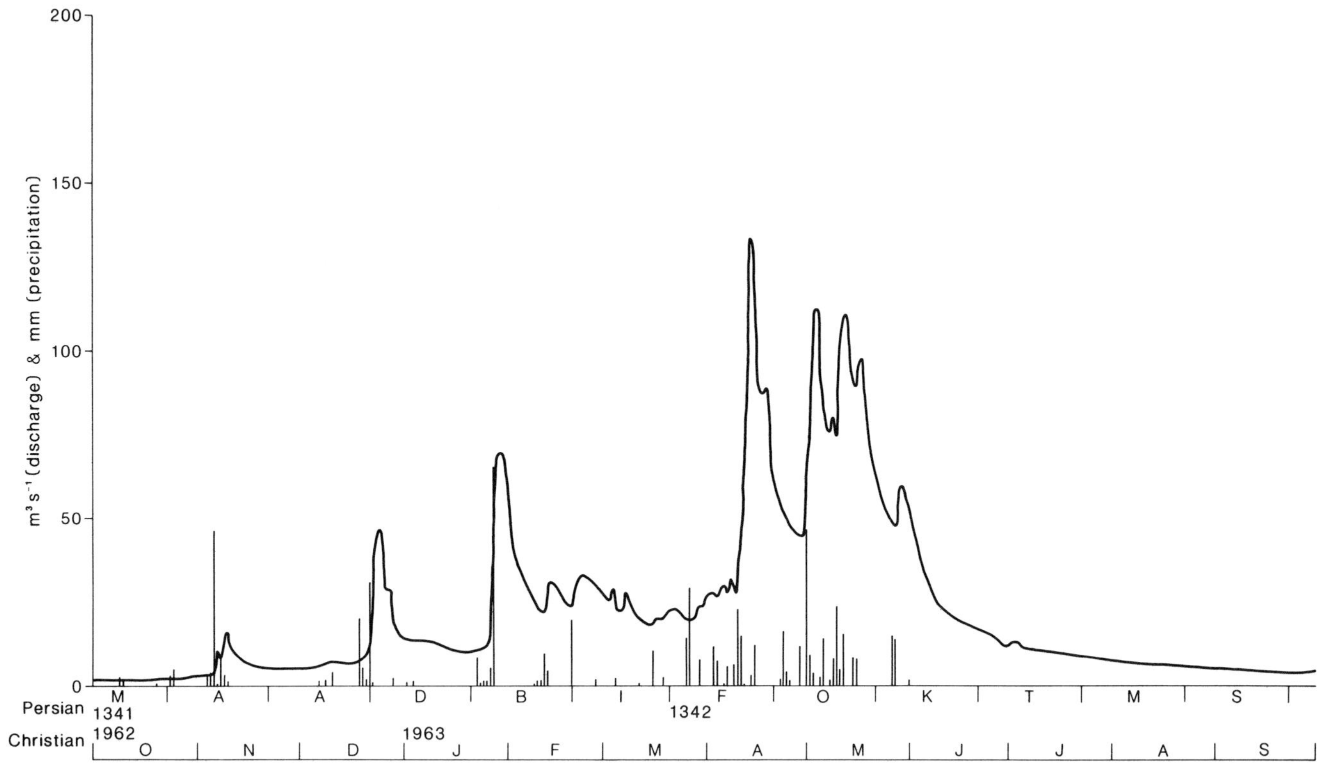
m³s⁻¹ (discharge) & mm (precipitation)
200
150
100
50
0
M
A
A
D
B
I
F
O
K
T
M
S
Persian
1341
1342
Christian
1962
1963
O
N
D
J
F
M
A
M
J
J
A
S

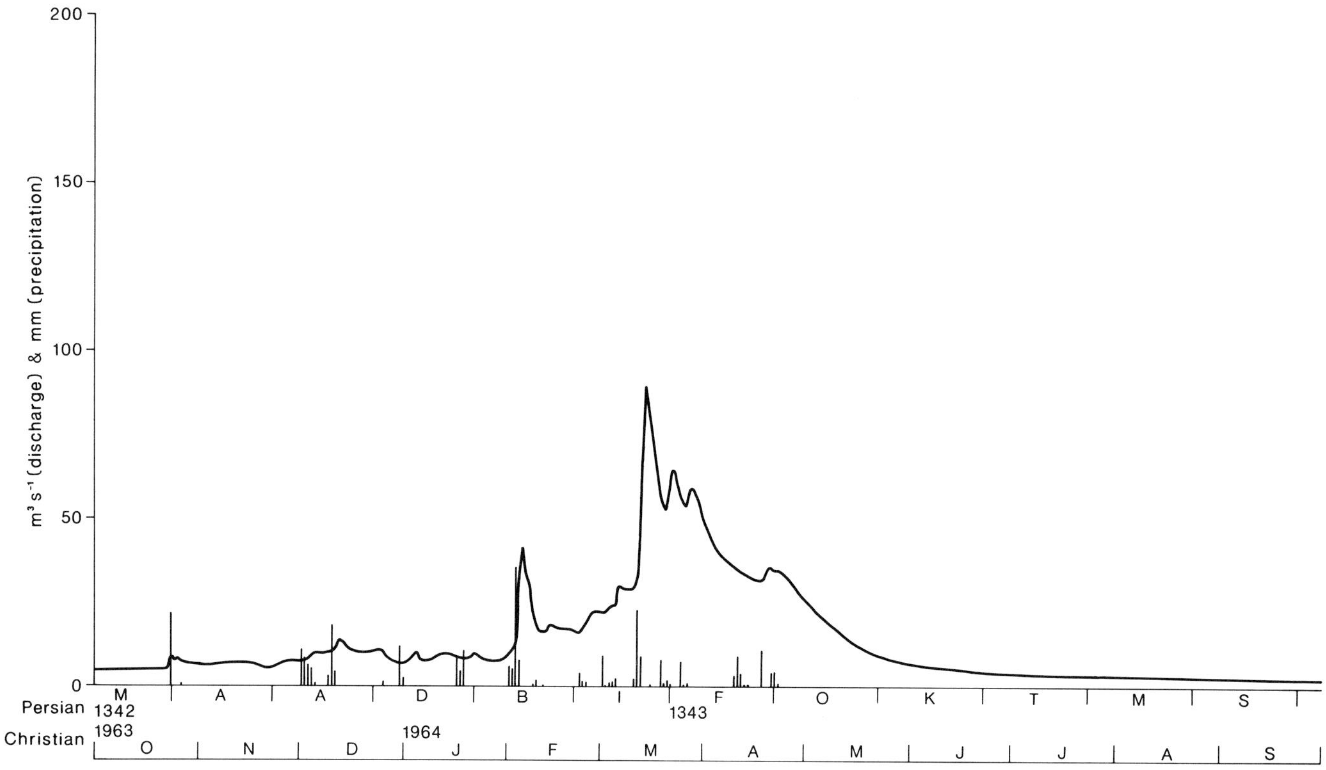
200
150
100
50
0
m³s⁻¹ (discharge) & mm (precipitation)
Persian
M
1342
A
A
D
B
I
F
1343
O
K
T
M
S
Christian
1963
1964
O
N
D
J
F
M
A
M
J
J
A
S

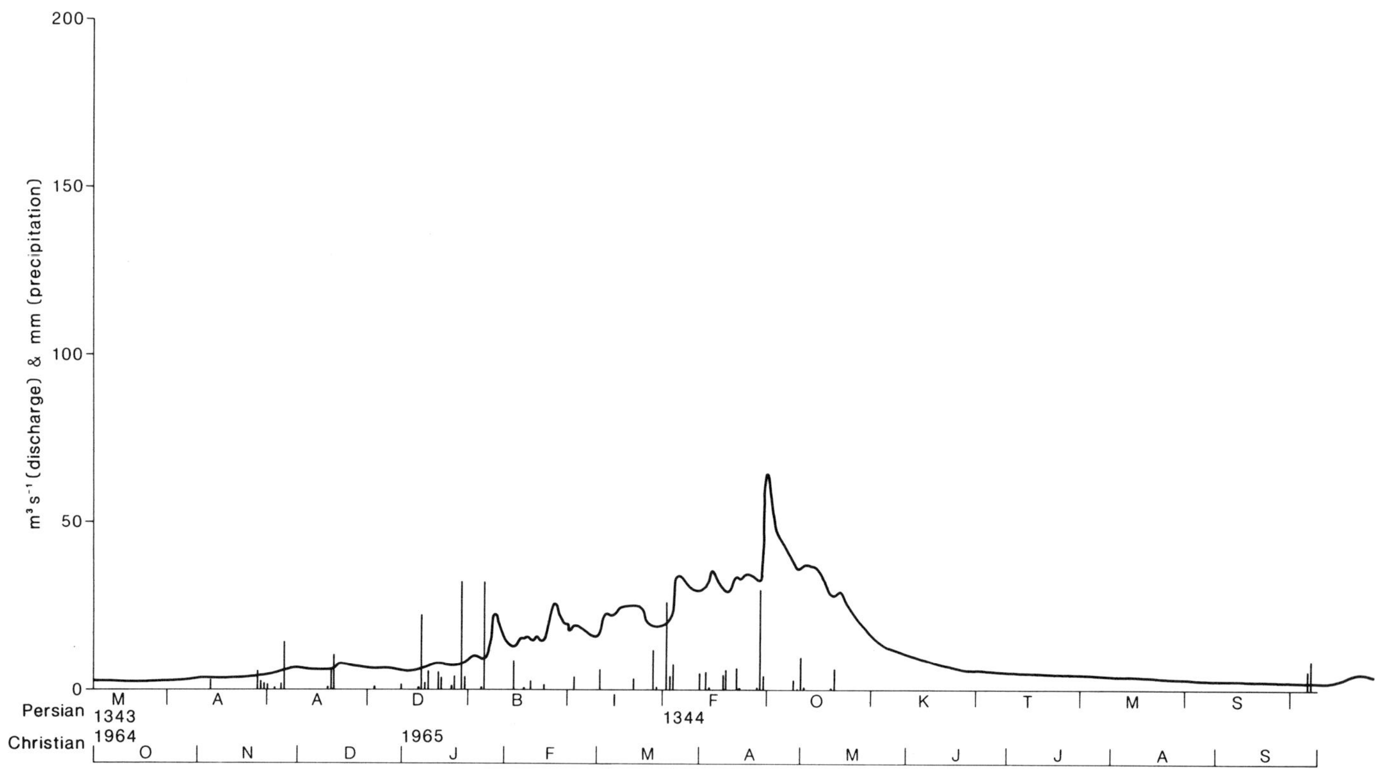
200
150
100
50
0
m³s⁻¹ (discharge) & mm (precipitation)
Persian
M
1343
A
A
D
B
I
F
1344
O
K
T
M
S
Christian
1964
1965
O
N
D
J
F
M
A
M
J
J
A
S

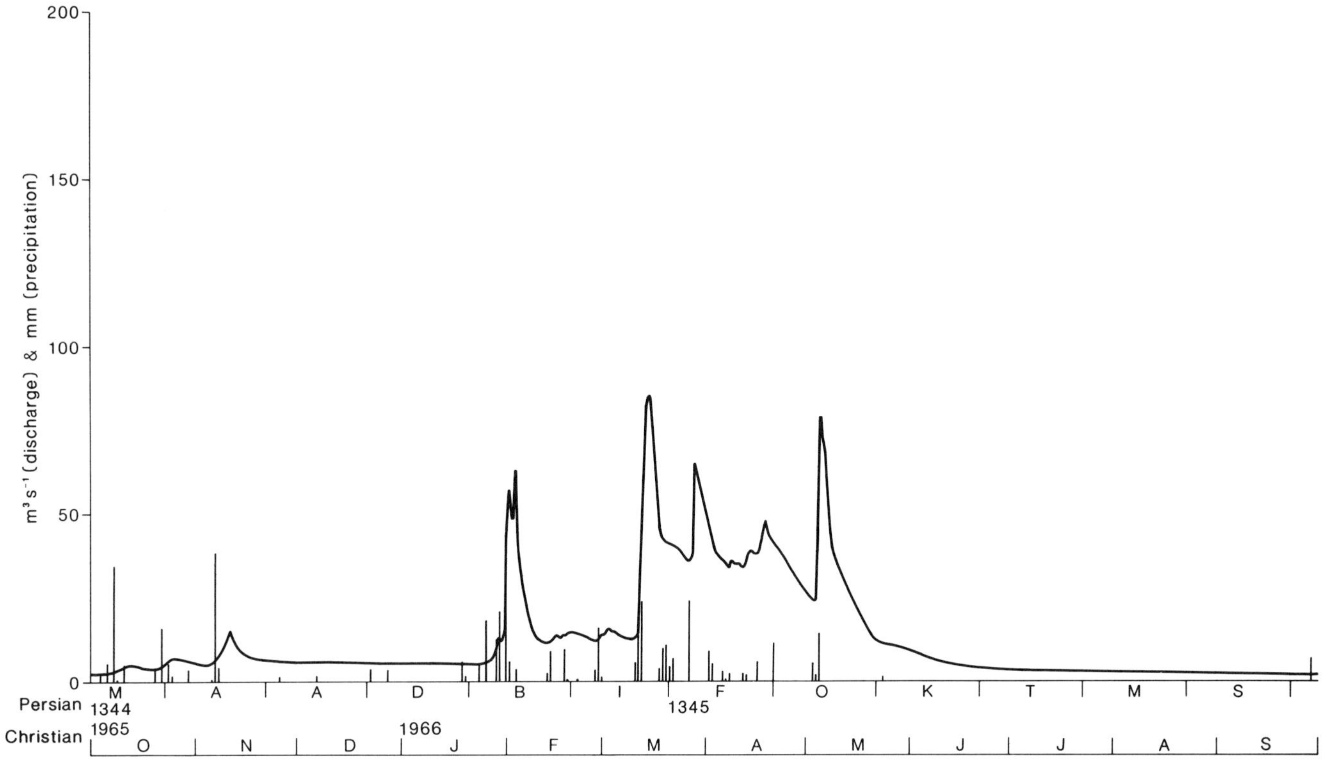
200
150
100
50
0
m³s⁻¹ (discharge) & mm (precipitation)
M
A
A
D
B
I
F
O
K
T
M
S
Persian 1344
1345
Christian 1965
1966
O
N
D
J
F
M
A
M
J
J
A
S

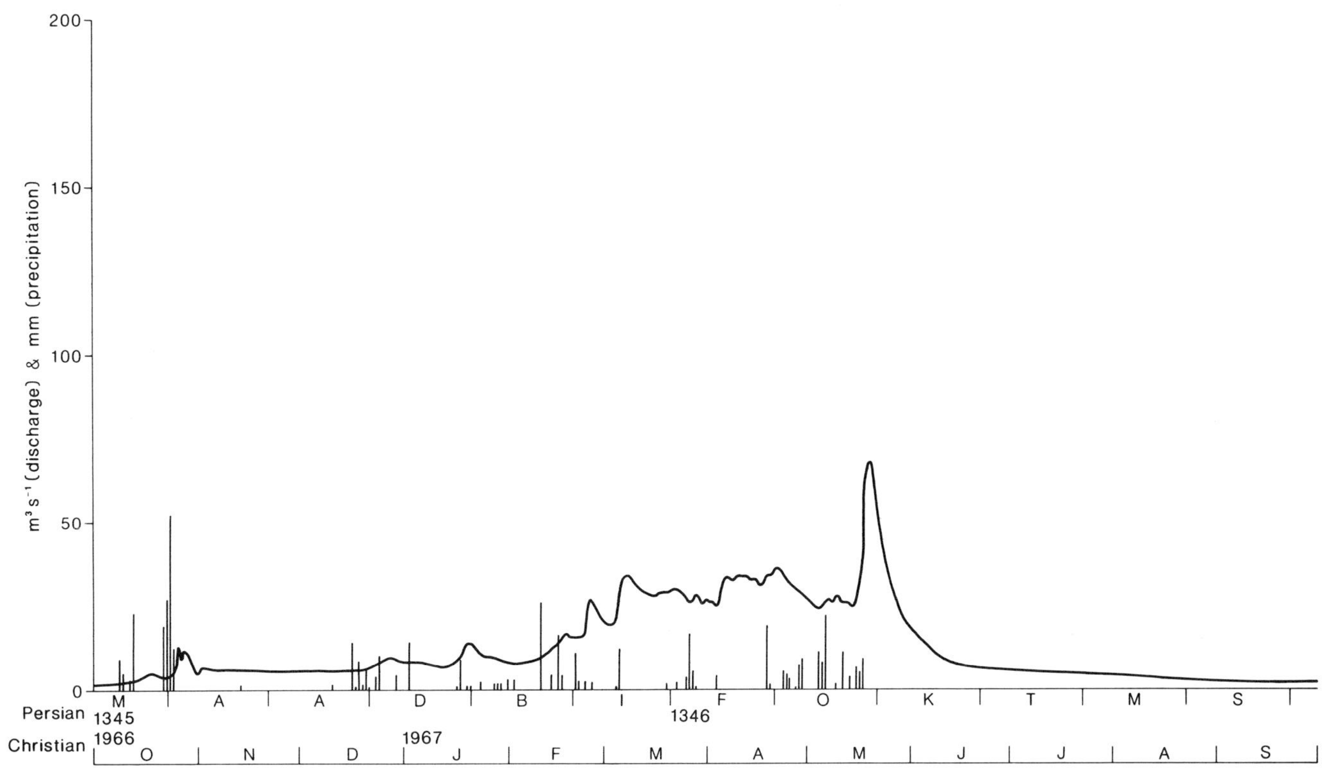
200
150
100
50
0
m³s⁻¹ (discharge) & mm (precipitation)
M
A
A
D
B
I
F
O
K
T
M
S
Persian 1345
1346
Christian 1966
1967
O
N
D
J
F
M
A
M
J
J
A
S

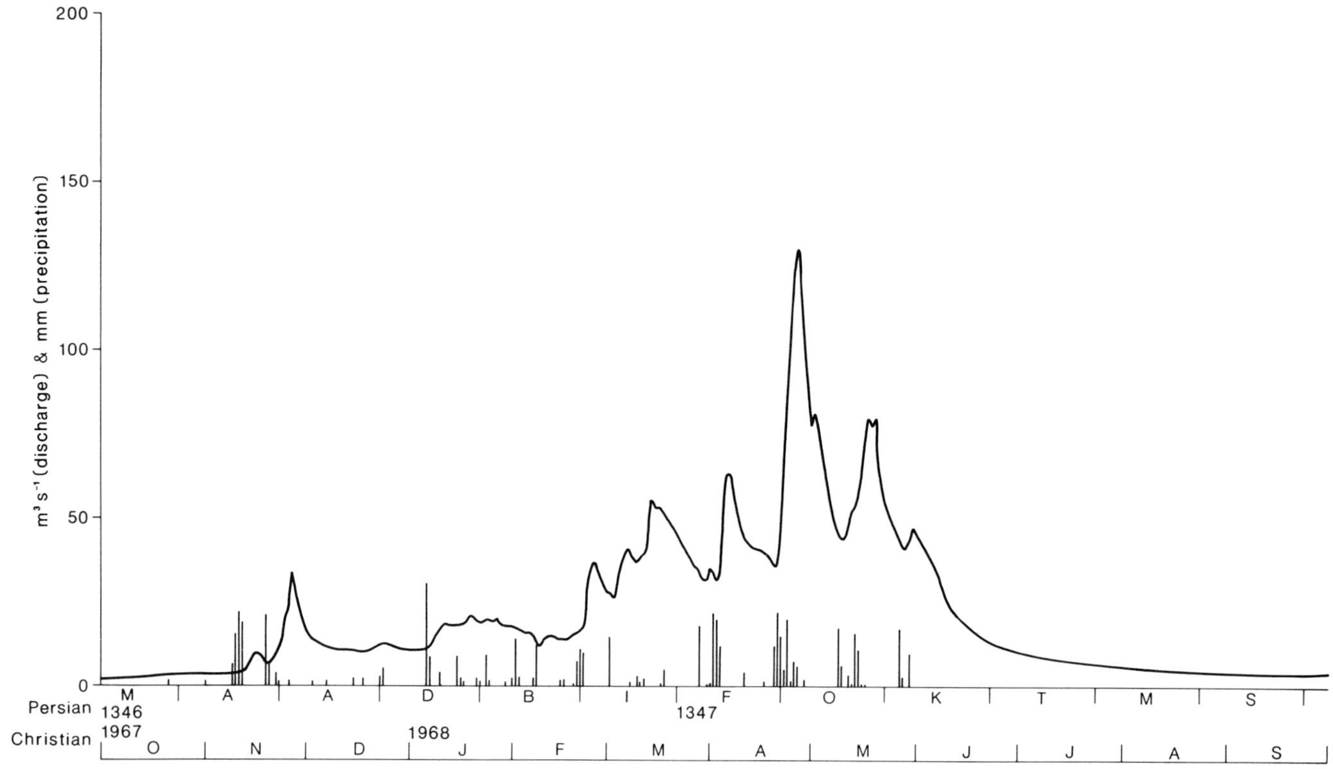
200
150
100
50
0
m³ s⁻¹ (discharge) & mm (precipitation)
Persian
M A A D B I F O K T M S
1346
1347
Christian
1967
1968
O N D J F M A M J J A S

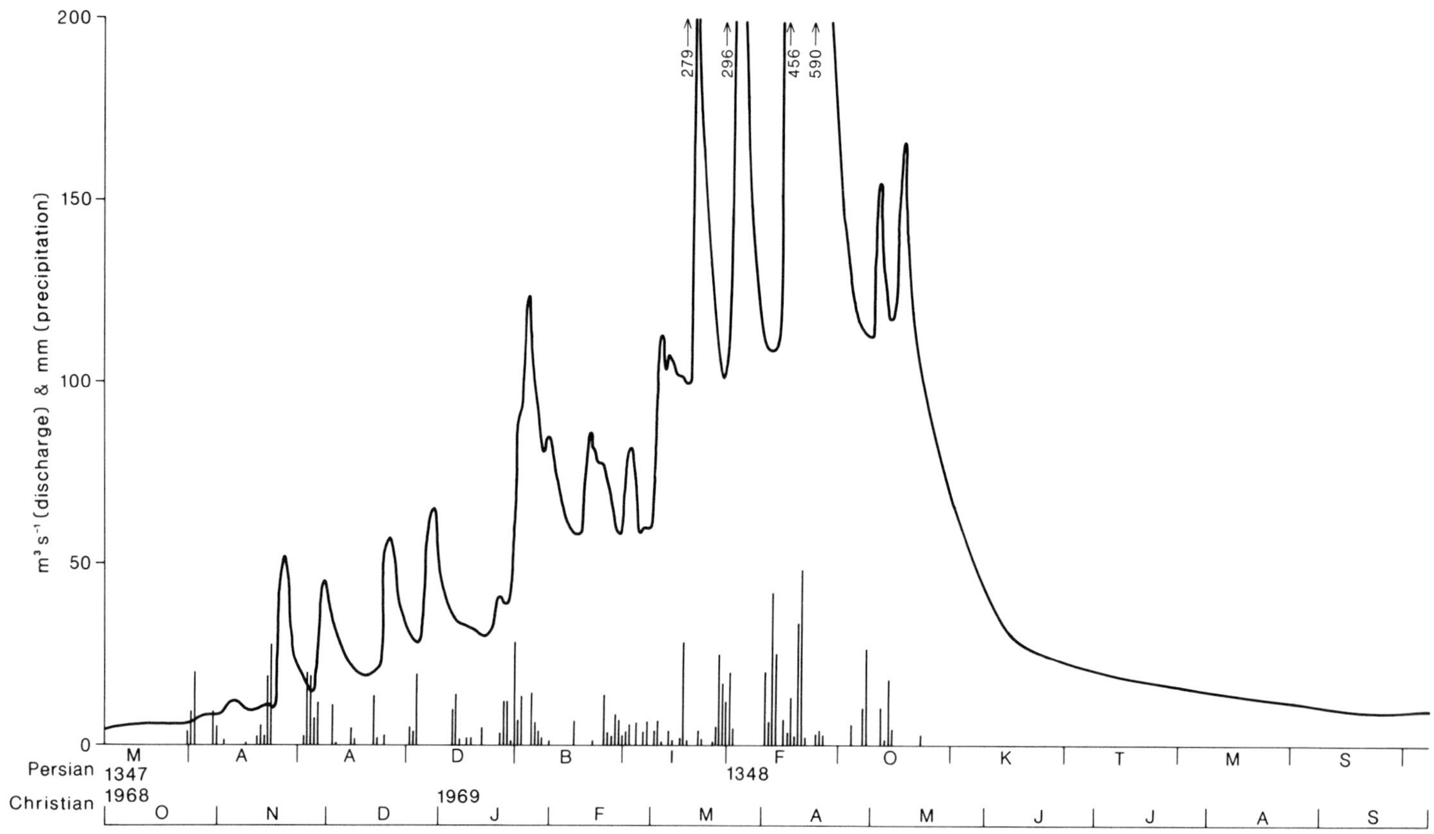
200
150
100
50
0
m³ s⁻¹ (discharge) & mm (precipitation)
279
296
456
590
Persian
M A A D B I F O K T M S
1347
1348
Christian
1968
1969
O N D J F M A M J J A S

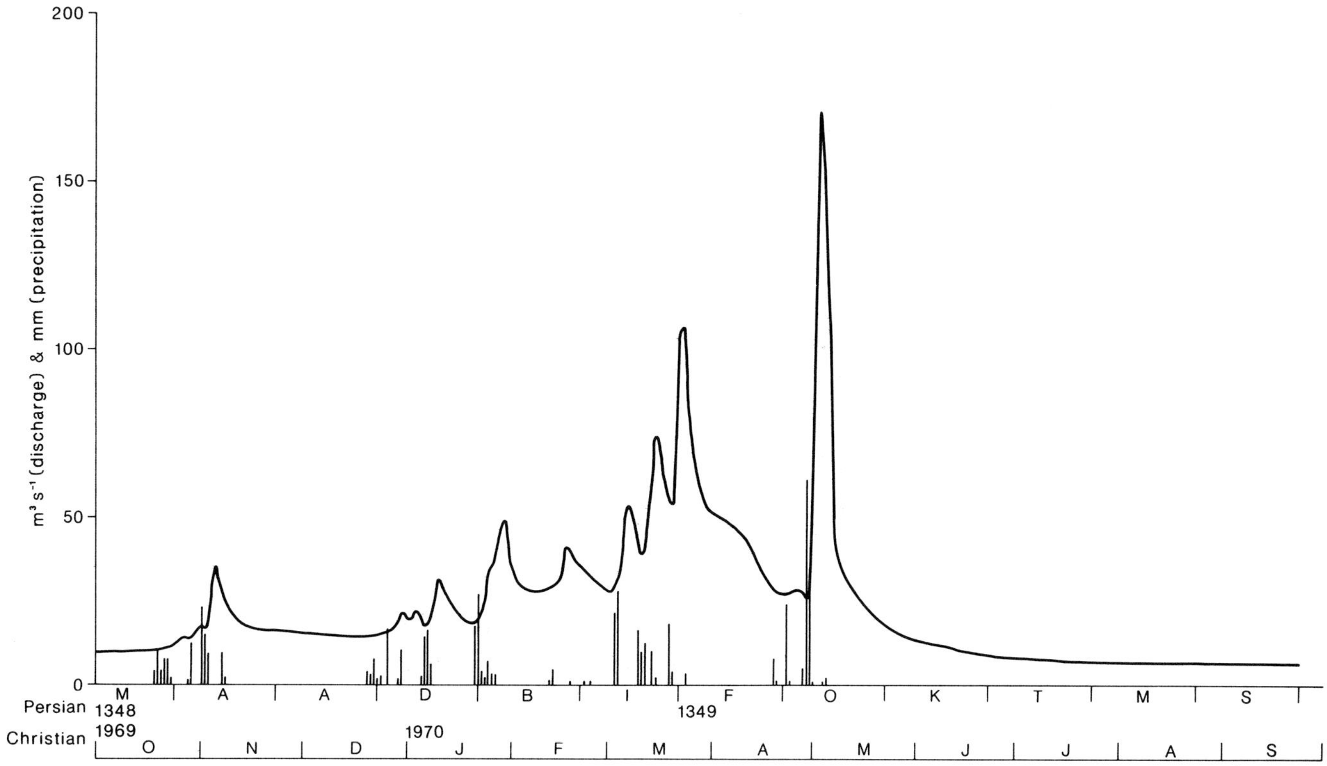
200
150
100
50
0
m³ s⁻¹ (discharge) & mm (precipitation)
Persian
M
1348
A
A
D
B
I
F
1349
O
K
T
M
S
Christian
1969
O
N
D
1970
J
F
M
A
M
J
J
A
S

References

AB-AV Khakh, 1973. *Complementary Investigations and Feasibility of Groundwater Development*. Kermanshah—West Shahabad Project, vol. 1. Tehran, Iran Ministry of Energy, Water Division.

Adams, R. Mc., 1965. *Land Behind Baghdad*. Chicago, University of Chicago Press.

Brichambaut, Ch. P. and Wallén, C. C., 1962. *Technical Report on a Study of Agroclimatology in Semi-Arid and Arid Zones of the Near East*. Paris, United Nations, Food and Agricultural Organization, and UNESCO, World Meteorological Organization.

Brookes, I. A., 1982. Geomorphological evidence for climatic change in Iran during the last 20,000 years. *In* J. L. Bintliff and W. van Zeist (eds.), *Palaeoclimates, Palaeoenvironments and Human Communities in the Eastern Mediterranean Region in Later Prehistory*, pp. 191–228. British Archaeological Reports, International Series, I133 (i and ii).

Bull, W. B., 1978. The alluvial fan environment. *Progress in Physical Geography* 1:222–270.

Costa, J. E., 1978a. Colorado Big Thompson flood: geologic evidence of a rare hydrologic event. *Geology* 6:617–620.

——, 1978b. Holocene stratigraphy in flood frequency analysis. *Water Resources Research* 14:626–632.

Danin, A., Gerson, R. S., and Garty, J., 1983. Weathering patterns on hard limestone and dolomite by endolithic lichens and cyanobacteria: supporting evidence for eolian contribution to terra rossa soil. *Soil Science* 136(4): 213–217.

Dewey, J. F., Pitman, W. C., III, Ryan, W. B. F., and Bonnin, J., 1973. Plate tectonics and the evolution of the Alpine system. *Bulletin of the Geological Society of America* 84:3137–3180.

Finnie, R. A., 1979. The influence of geomorphology on archaeological site distributions in the Mahidasht, Central West Iran. B.A. thesis, Department of Geography, York University, Toronto, 104 pp.

Frolking, T. A., Jackson, M. L., and Knox, J. C., 1983. Origin of red clay in the loess-covered Wisconsin driftless uplands. *Soil Science Society of America Journal* 47:817–820.

Ganji, M. H., 1955. Les climats de l'Iran. *Bulletin de la Société géographique française* 28:195–299.

——, 1965. *Climatic Atlas of Iran*. Tehran, Institute of Geography, University of Teheran.

——, 1968. Climate. *In* W. B. Fisher (ed.), *The Land of Iran*. The Cambridge History of Iran, vol. 1, pp. 212–249. Cambridge, Cambridge University Press.

Gardner, R. M., 1977. Evidence concerning the existence of loess deposits at Tell Fara, Northern Negev, Israel. *Journal of Archaeological Science* 4:377–386.

Goudie, A. S., 1983. Calcrete. *In* A. S. Goudie and K. Pye (eds.), *Chemical Sediments and Geomorphology*, pp. 93–131. New York, Academic Press.

Hooke, R. LeB., 1968. Steady-state relationships on arid-region alluvial fans in closed basins. *American Journal of Science* 266:609–629.

Iran Ministry of Water and Power, 1956–1972. *Meteorological Yearbooks*. Tehran.

——, 1958–1970. *Hydrographic Yearbooks*. Tehran.

Issar, A. S. and Bruins, H. J., 1983. Special climatological conditions in the deserts of Sinai and the Negev during the latest Pleistocene. *Palaeogeography, Palaeoclimatology, Palaeoecology* 43(1–2):63–72.

Keith, M. L. and Anderson, G. M., 1963. Radiocarbon dating: fictitious results with mollusk shells. *Science* 141: 634–637.

Kirkby, M. J., 1977. The land and water resources of the Deh Luran and Khuzistan Plains. Appendix 1. *In* F. Hole, *Studies in the Archaeological History of the Deh Luran Plain: The Excavation of Chagha Sefid*. Memoirs of the Museum of Anthropology, University of Michigan, no. 9, pp.251–288. Ann Arbor, University of Michigan.

Kirkby, M. J. and Kirkby, A. V., 1976. Geomorphic processes and the surface survey of archaeological sites in semi-arid areas. *In* D. A. Davidson and M. L. Shackley (eds.), *Geoarchaeology—Earth Science and the Past*, pp. 229–253. London, Duckworth's.

Lambton, A. K. S., 1980. Kirmanshah. In *The Encyclopedia of Islam*, pp. 169–170. Leiden, E. J. Brill.

Leeder, M. R., 1975. Pedogenic carbonates and flood sediment accretion rates: a quantitative model for arid-zone lithofacies. *Geological Magazine* 112(3):257–270.

Macleod, D. A., 1980. The origin of the red Mediterranean soils in Epirus, Greece. *Journal of Soil Science* 31(1): 125–136.

Maire, R., 1978. Première reconnaissance sur un grand karst d'altitude des Zagros: le plateau de Ravansar, Province de Kermanshah, Loristan, Iran. *Bulletin de l'Association des Géographes Français*, no. 449, pp. 51–58.

Netterberg, F., 1980. Geology of South African calcretes I: terminology, description, macrofeatures, and classification. *Transactions of the Geological Society of South Africa* 83: 255–283.

Passega, R., 1964. Grain size representation by CM patterns as a geological tool. *Journal of Sedimentary Petrology* 34:830–847.

Petroshevsky, I. P., 1968. The socio-economic conditions of Iran under the Il-Khans. *In* J. A. Boyle (ed.), *The Saljuq and Mongol Periods*. The Cambridge History of Iran, vol. 5, pp. 483–637. Cambridge, Cambridge University Press.

Rathjens, C., 1965. Ein Beitrag zur Frage der Solifluktionsgrenze in den Gerbirge Vorderasiens. *Zeitschrift für Geomorphologie* 9:35–49.

Raynal, R., 1977. Étude comparé en altitude des processus periglaciaires actuels dans les hauts massifs du Maroc et de l'Iran. *Akademische Wissenschaftlichen Göttingen, Mathematische—Physische Klasse, Abhandlungen,* ser. 3, vol. 31, pp. 275–289.

Rieben, H., 1955. The geology of the Teheran Plain. *American Journal of Science* 253:617–639.

Singer, A., 1967. Mineralogy of the non-clay fraction from basaltic soils in the Galilee, Israel. *Israel Journal of Earth Sciences* 16:215–228.

Stocklin, J. and Nabavi, M. H., 1972. *Tectonic Map of Iran.* Tehran, Iran Geological Survey.

Takahashi, K. and Arakawa, H., 1981. *Climates of Southern and Western Asia.* World Survey of Climatology, vol. 9. New York, Elsevier.

Takin, M., 1972. Iranian geology and continental drift in the Middle East. *Nature* 235:147–150.

van Zeist, W., 1967. Late Quaternary vegetation history of Western Iran. *Review of Palaeobotany and Palynology* 2:301–311.

van Zeist, W. and Bottema, S., 1977. Palynological investigations in Western Iran. *Palaeohistoria* 19:19–85.

Vita-Finzi, C., 1969. Late Quaternary alluvial chronology of Iran. *Geologische Rundschau* 58(3):951–973.

Waltham, A. C. and Ede, D. P., 1973. The karst of the Kuh-e Parau. *Transactions of the Cave Research Group, Great Britain* 15:27–40.

Woillard, G., 1978. Grande Pile peat bog: a continuous pollen record for the last 140,000 years. *Quaternary Research* 9(1):1–21.

Wolman, M. G. and Leopold, L. B., 1957. Flood Plains. *United States Geological Survey,* Professional Paper 282-C, pp. 87–107.

Wright, H. E., Jr., 1968. Natural environment of early food production north of Mesopotamia. *Science* 161:334–339.

Yaalon, D. H. and Ganor, E., 1973. The influence of dust on soils during the Quaternary. *Soil Science* 116(3):146–155.

——, 1975. Rate of eolian dust accretion in the Mediterranean and desert fringe environments of Israel. *Proceedings 10th International Sedimentological Congress, Nice,* vol. 2, pp. 169–174.

Zohary, M., 1973. *Geobotanical Foundations of the Middle East.* 2 vols. Geobotanica Selecta 3. Stuttgart, G. Fischer, and Amsterdam, Swets & Zeitlinger.

Plates

PLATE 1. Geology of the Mahidasht Project area, with (inset) southwest-northeast geological section along A–A′.

PLATE 2. Hydrography of the Mahidasht Project area, showing perennial and intermittent streams, springs, and gauging stations on Qara Su at Doab (D), Pol-i Khoneh (K), and Gharbaghestan (G).

PLATE 5. Distribution and height of archaeological sites containing pre–Bronze Age material in the Mahidasht Project area. Also shown: extent of alluvium, major towns and roads, 10 km × 10 km UTM grid. Sites in unsurveyed squares were located on line traverses.

(Plates 3 and 4 in envelope.)

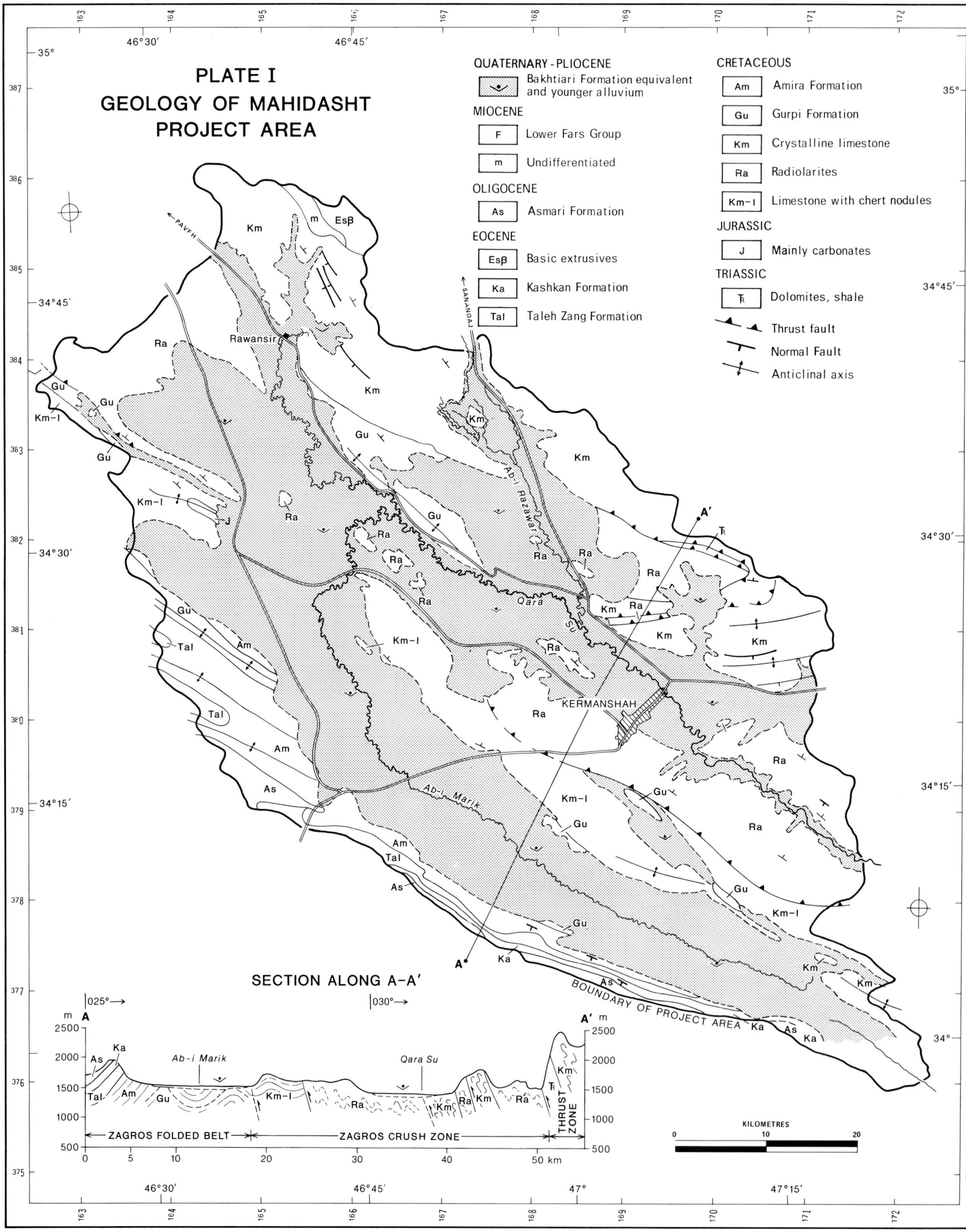

SOURCE: U.S.ARMY SERVICE, 1:250,000, EDITION 1 (1958). SHEETS N138-7.

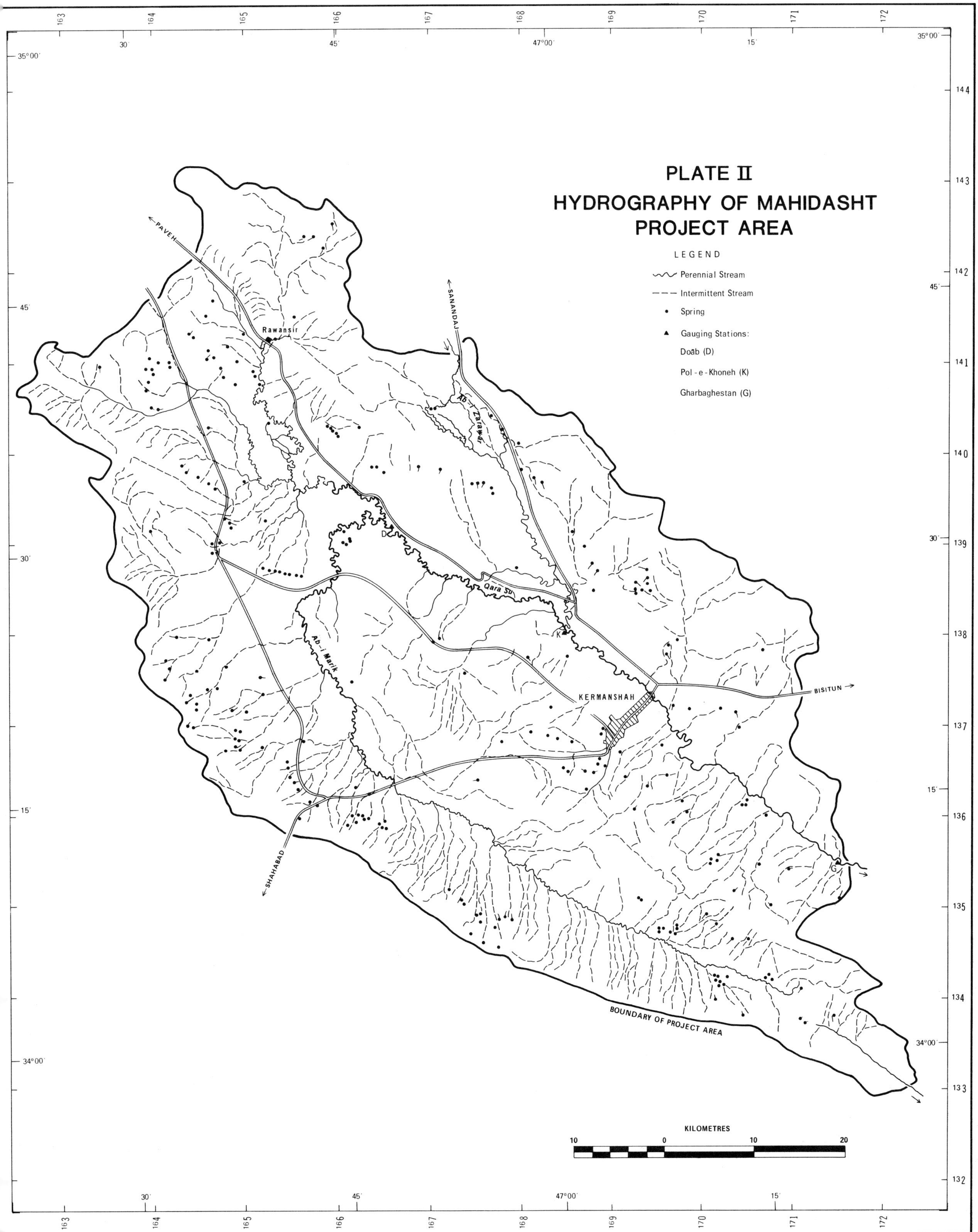

SOURCE: U.S. ARMY SERVICE, 1:250,000, EDITION 1 (1958), SHEETS N138-7,8,12

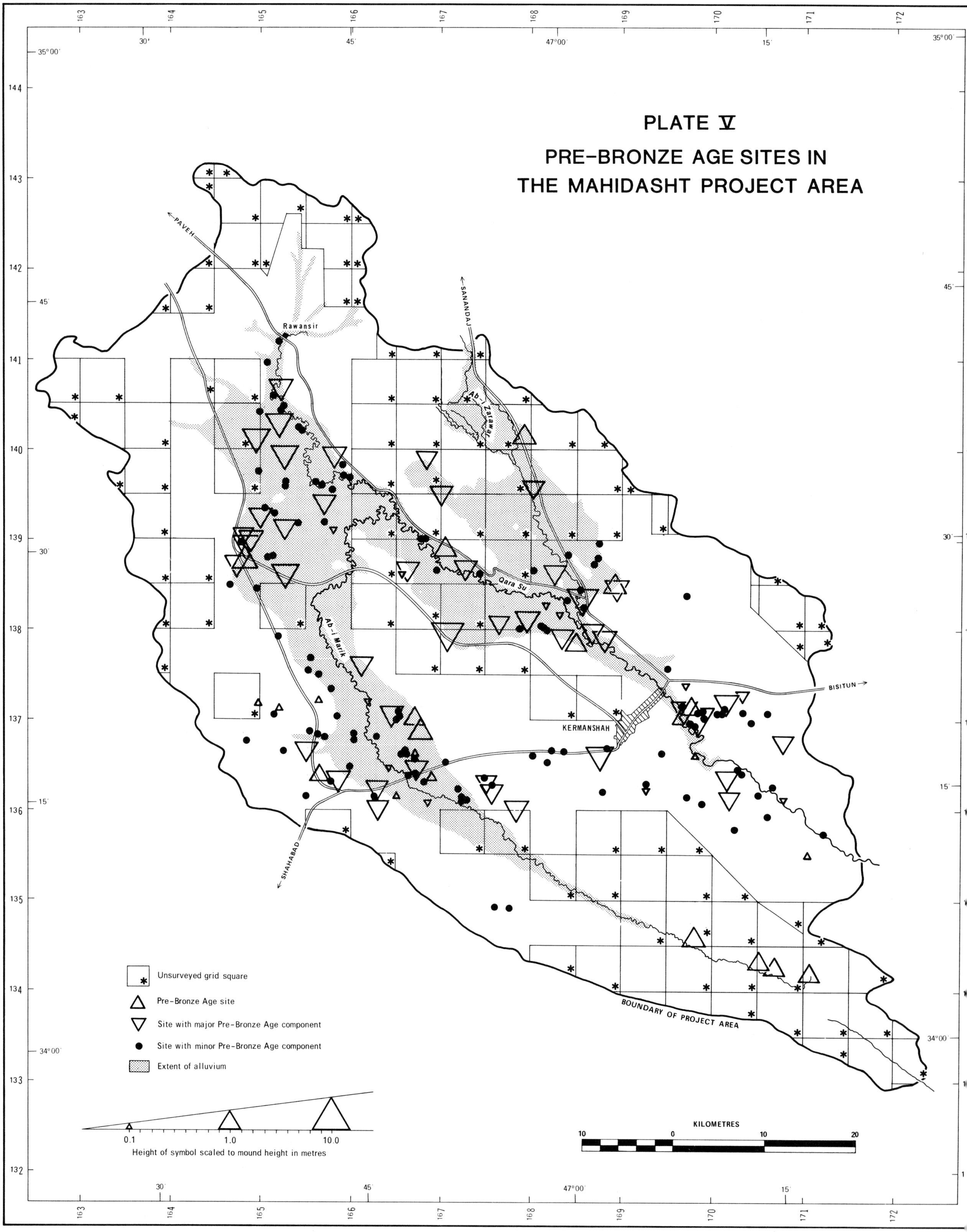

SOURCE: U.S. ARMY SERVICE, 1:250,000, EDITION 1 (1958), SHEETS N138-7